A TIME TO BE BORN

A Time to Be Born

An Almanac of Animal
Courtship and Parenting

text by Lorus and Margery Milne
illustrations by Sarah Landry

SIERRA CLUB BOOKS   SAN FRANCISCO

The Sierra Club, founded in 1892 by John Muir, has devoted itself to the study and protection of the earth's scenic and ecological resources-mountains, wetlands, woodlands, wild shores and rivers, deserts and plains. The publishing program of the Sierra Club offers books to the public as a non-profit educational service in the hope that they may enlarge the public's understanding of the Club's basic concerns. The point of view expressed in each book, however, does not neccessarily represent that of the club. The Sierra Club has some fifty chapters coast to coast, in Canada, Hawaii, and Alaska. For information about how you may participate in its programs to preserve wilderness and the quality of life, please address inquiries to Sierra Club, 530 Bush Street, San Francisco, CA 94108.

Library of Congress Cataloging in Publication Data

Milne, Lorus Johnson, 1912–
  A time to be born.

  Includes index.
  1. Mammals—Reproduction. 2. Mammals—Behavior.
I. Milne, Margery Joan Greene, 1914–   . II. Title.
QL739.3.M54   599.05′6   82–3300
ISBN 0–87156–317–7   AACR2

Book design by Leigh McLellan
Illustrations by Sarah Landry

Printed in the United States of America

10 9 8 7 6 5 4 3 2 1

Contents

Sex, Season, and Survival

FOR MANY OF the world's mammals, the season of birth is critical for survival. Any animal, young or adult, may have difficulty finding enough shelter and food when winter spreads its long nights and chill upon the land. Neither a mother encumbered by her unborn young nor newborn mammals can compete well when the struggle for limited resources rises to its annual peak.

Reduced rainfall, for example, and lower temperatures at higher latitudes make less water available to green plants, which constitute the basic food resource in every land. Less water means less new growth in plants, and therefore less food for animals. Family behavior directly corresponds to these variations in the availability of food. Charles Darwin and Alfred Russel Wallace recognized these facets of the wild environment in developing their theory of natural selection. New lives are snuffed out whenever inherited behavior fails to schedule sexual activity and season of birth for the best periods of the year; selection simply tailors inherited behavior toward survival.

Wild animals react far more than we do to seasonal changes in climate. Although people talk about the weather almost daily, we are relatively immune to the ordinary challenges presented by climate. Our remote ancestors probably responded to inherited, pre-conscious urges with survival behaviors comparable to those of the food-storing mammals. Gradually they evolved an increased awareness of their surroundings and a learned understanding of what to do at each season of the year. Now we regard ourselves as equally competent in any month.

The hour of day and month of birth hold important places among the specifications that members of every species inherit so inconspicuously. Hidden away in the double-helix molecules of DNA are the master templates for all chemical processes (including those that form and maintain body structures for life) and also the script for their implementation, which—like stage directions—spell out the sequence and time for each event. The DNA molecules remain within the chromosomes, inside the nucleus of each active cell, at least until they are firmly committed to some one role in the individual. Yet they are not walled off within an impregnable fortress. They can be influenced by the environment, through nutrition, through length of night, and through messages from outside the body. The environment provides the master clock for sexual activity in particular. Inherited nature and environment-controlled nurture are linked components of the feedback loops that sustain and schedule life. Their level of complexity exceeds that of any computer our human ingenuity is likely to devise.

The specifications for each species define what foods the animal can use at every age, what shelter and other amenities it needs from the nonliving environment, what messages it can send and receive, what course of action or of inaction will follow in response to challenges, and when it will be ready to start a family. The genetic heritage prescribes the ecological niche into which the individual fits, as long as it encounters roughly the same opportunities its ancestors met.

Scientific knowledge can seldom explain the detailed differences in sexual behavior that characterize each kind of mammal. We do not know why a newborn opossum should be the size of a honeybee and emerge from its mother's birth canal to make its way unaided to a milk-filled nipple in her pouch. This first journey comes after a pregnancy lasting only thirteen days. It interrupts a program of growth that can produce a full-grown mammal the size of a large house cat. We can only observe that the tradition inherited by the opossum is shown by all marsupial (pouched) mammals. Uniformly they produce young at an extraordinarily early stage of development, before eyes, ears, hind legs, tail, or body hair are evident. We can appreciate that this allows a female marsupial to divest herself of her offspring more efficiently and safely than a pregnant placental mammal could in times of unusual food shortage. Environmental uncertainty seems a commoner challenge in modern Australia, where marsupials developed such diversity, than we might expect it to have been in tropical South America millennia ago, where the marsupials originated.

It takes a mother cat longer—a full 42 days of pregnancy—to produce a kitten with all appendages and fur, even though the newborn kitten's eyes are sealed shut and it shows no indication of hearing any sound. The opossum and the cat, although both are mammals, carry on different developmental programs with much the same result. Rarely in nature is any one heritage so superior to all others that it becomes the sole system for survival.

Life survives by compromises. It cannot go on indefinitely following the old inherited ways. It cannot endure by reacting fully to each vagary of climatic change. Only by compromising can life accommodate itself to a chronically unstable world, where mountains are thrust up and erode down, seas invade the continents and drain away or evaporate to salt flats, whole oceans expand while continental masses drift apart, and new kinds of plants and animals replace more ancient types. All these changes show great local variation in rate and extent. The inherited schedule of family life must adapt by compromise, for uniformity offers no route to the future. The benefits that the mammal derives from following a particular schedule may be a well-hidden secret in any one generation, and still allow a cumulative gain over the decades and centuries.

Each creature unwittingly invests its resources in reproducing itself, and finds many ways to reduce the risks. Timing may be the easiest and simplest to adjust according to the success it brings. An almanac of mating and birthing among mammals reveals the strategy that leads later to encounters with seasonal change in a pattern that favors survival. The almanac can start with the month in which spring arrives in the Northern Hemisphere and autumn south of the equator. For many a mammal, this is the preferred time to begin life.

CHAPTER ONE

The Mad Month of March

ERRATIC WEATHER IS so characteristic of the weeks before and immediately after equinox in March that choosing a particular date then to make a field trip in northeastern parts of America or northern Europe provides a major element of risk. The season seems too early for migratory birds to arrive from sunnier lands to the south, or for local mammals to explore widely in search of mates. Yet survival of young in later months is often served by ventures begun in March.

On a blustery March day a few years ago, a traveler through New Hampshire found his way to the University's Zoology Department to inquire about a strange animal he had unintentionally run over on the highway close to town. He thought it was a fox as it bounded across the road, until it paused inopportunely just ahead of his fast-moving vehicle. We identified the cooling body for him as the largest of the weasel tribe —a fisher *(Martes pennanti)*, known variously as fisher cat and pekan. Probably it ended up so close to town while searching from one patch of woodland to another for a mate.

The Fishers that Never Fish

Weighing close to eleven pounds, the fisher was a handsome male with thick, soft, blackish brown fur. Its nose and ears were shorter than those of a fox, its tail less bushy, and its paws more obviously armed with claws. These claws are necessary for tree climbing, an important skill for fishers. This strenuous activity takes the animal to its favorite prey: porcupines, raccoons, red and gray squirrels, and sometimes a martin—a smaller member of the tree-traveling weasel tribe. On the ground, a fisher takes rabbits and hares, woodchucks and marmots, mice and chipmunks, and even an occasional fox or lynx that is unprepared for a sudden attack. Some people credit the fisher with being able to dispatch a full-grown deer by biting through its throat and cutting the jugular vein. Despite its name, a fisher does not go fishing, although it is willing to retrieve a dead or dying fish as big as a salmon and make a meal of it. The name fisher may well have arisen through confusion with the much smaller, fish-catching mink.

Female fishers, which are somewhat smaller than the males and have a silkier fur, seem to keep aware of suitable tree holes they find, often as the final futile refuge of a red squirrel they have pursued and eaten. Young fishers are born in March or early April in one of these hideaways. About a week later the mother will be receptive to a mate—a condition she maintains for only a day or two. It is up to some male to detect her special scent upon the equinoctial winds and to wait for her outside her tree hole. She will attack him furiously if he tries to enter. Even in the open, her tolerance for him is unpredictable and brief, requiring him to bide his time until hers has come. Thereafter, for about 51 weeks, she will be pregnant. Yet her young do not develop continuously during all those weeks. Almost like the embryo in a seed, they progress through early stages and then pause. They remain dormant inside her womb through late spring and summer, recommencing their growth only in autumn, in time to be born into the drafty, chilly world of March.

Newborn fishers are blind and helpless at first. Their eyes remain closed until about seven weeks of age, and the young animals are unready to venture beyond the den until June at the earliest. Then they begin accompanying their mother on hunting trips and sampling her catch immediately, without having to wait for her to bring it home. By late autumn, the little family disperses. Only the most precocious of the young males will be ready to look for a mate the following spring. The other males will wait another year, surviving on nuts, fruits, and less nourishing plant material if winter prey is scarce.

Legislators in New Hampshire heard about fishers on two separate occasions recently. The first came while they were debating at great length whether to authorize foresters and orchardists to roll poisoned apples into the winter dens of porcupines. Economic loss was widespread because the quill-bearers were emerging periodically in unprecedented numbers to feast on the nutritious inner bark of trees, seriously defacing some trees and causing the death of others by girdling. How could the legislators be guaranteed that no porcupine would bring a poisoned apple into the sunshine and discard the fruit almost undamaged, where a child, a horse or a deer might find it and become an unintended victim? Before any of the proposed new laws were passed or permits were issued for the control of porcupines, the need for human intervention disappeared. The population of quill-bearers began to shrink due to an unexpected increase in the number of fishers in the forests. The few fishers that had survived from earlier efforts by fur trappers benefitted from the extra supply of live porcupines. Raising record broods of young, the old fishers taught the juveniles how to flip a porcupine onto its back and attack through the unprotected belly surface. Foresters and orchardists were delighted to learn that they had a natural ally, particularly a nocturnal animal and one that would do its work without cost. The legislators relaxed too.

Then came a new request. If fishers were becoming numerous again, why not strike them from the list of protected animals? It might be possible to earn a bonus from the forest by trapping fishers for their pelts. After all, at the turn of the century, prime fisher skins sold for as much as $75 to $100. The fact that these record prices were paid for unusually large pelts from females only, and that each female skinned cost the fisher population the lives of her nurslings in a tree hole plus the embryos in her womb, was never mentioned. This was the reason that the previous work of trappers had so quickly brought fisher numbers close to the vanishing point.

Now sentiment is turning away from harvesting any wild animal for its fur unless it is a far-off, carefully managed operation, as is the case with the Pacific fur seals in Alaska. So fishers again have a chance to replenish their population, not only in New Hampshire but across America from Labrador to British Columbia, south in the mountains of California and in the Appalachians as well. March is the best time to find the tracks of the ardent males, or to see one abroad by day on mating business that cannot wait.

The Much-maligned Wolves of the North

Unlike the generally silent fisher, the gray wolf (*Canis lupus*) enlivens the night by voicing calls of many kinds. Raising its nose toward the sky, it tries a few short howls, then utters a long one whose pitch holds firm a while before descending toward a melancholy end: "Wup-wup-wup-wup-wup-WOOOOOOOOooooo." Anyone who hears this distinctive call stops to pay attention, for it sounds close at hand even when the wolf is a half mile away. Ordinarily other wolves join in, adding to the chorus while keeping to separate keys. Today it is a vocal signature of the wilderness, one that we are likely to encounter only by venturing into the Far North.

Three centuries ago, the calls of wild wolves could be heard almost anywhere in North America north of the twentieth parallel (in southern Mexico), and in Eurasia from Spain to Siberia. Now the wolf has vanished from most places where people live or try to raise domestic animals. Alaska may still have 5,000 wolves. More live in the Canadian Northwest Territories, perhaps 12,000 in forests of northern Ontario and a similar number in northern Quebec. These areas, and comparable wild country in Eurasia where some wolves exist, represent about 1 per cent of the former range of these powerful wild dogs.

Winter and earliest spring are the times when wolves become especially vocal. From January to March, their excitement mounts. Two-year-olds and some three-year-olds are discovering their sexual vigor for the first time, leading them to travel in search of mates beyond their normal boundaries. Pairs that have mated in previous years renew their courtship and prepare to raise another family. Since the period of gestation is only 63 days, just as in most kinds of domestic dogs, all this activity results in a wave of births beginning in late March and peaking in May. The timing differs little all the way from the Arctic to the tropics.

Depending mostly upon the food supply, a litter of wolves may be as small as two or as large as fourteen. The young that survive stay with the parents for two or three years and comprise the customary pack. Where hunting is good, packs sometimes join forces at least temporarily. But since the average wolf consumes more than seven pounds of meat daily in summer and more than eight pounds daily from October through May, prey must be abundant for a limited area to support a large number of wolves. On most ranges, each wolf needs about ten square miles of hunting territory. This is likely to mean that every square mile of it must sustain at least ten deer or their equivalent in food a wolf can use.

The former range of wolves in North America supplied an abundant variety of prey. Musk oxen and caribou in the Far North, moose and elk in the northern forests, deer farther south along the forest edges, bison and pronghorns on the plains, and mountain sheep and goats in the mountains were spaced out by snacks on smaller animals such as beaver, prairie dogs, hares and rabbits, lemmings, and mice. When settlers from Europe began introducing livestock, wolves adjusted so readily that federal aid was sought to exterminate the predators.

In only one small part of the United States have wolves found reasonable privacy and complete protection: Isle Royale in Lake Superior. Apparently one pack reached this sanctuary across the ice from Canada during a long cold period in the late 1940s. L. David Mech began observing their descendants while he was a graduate student at Purdue University. His census figures and those compiled by more recent investigators show a variation in the number of wolves from 18 to 28 by midwinter, when food is scarce. Of these animals, 15 to 16 associate in a single pack

that travels over the entire 210 square miles of the island. The other wolves, as loners and in twos and threes that seem not to breed, remain as remote as possible from the main pack and have difficulty finding enough to eat.

For all of the wolves on Isle Royale, the principal food resource is a moose herd that totals about 600 individuals each year just before the calving season begins. The herd remains almost constant because its natural increase is matched so well by the predation of the wolves. By selecting the young and the old moose, the wolves prevent the herd from exceeding its food supply, and leave as the breeding stock only the healthiest, strongest and most agile animals—making the hunt harder than ever.

To find prey, the hungry wolves trot tirelessly from late afternoon until early morning, traveling at about five miles per hour while remaining alert to sound, scent, and whatever their sharp eyes can see. This simple activity brings frequent reward on Isle Royale, for as Dr. Mech discovered in his attempts to follow the wolves, they average less than 23 miles between one kill and the next. This lets them concentrate their traveling into about 9 days each month, with about 22 days left to feed at leisure on the meat they have brought down.

Despite the ruthless vigor with which wolves attack their chosen prey, they normally avoid people who show any signs of life. Even in darkness, when the wolves are abroad, they do not pounce on campers asleep in the open. Hunters who have died of cold or starvation in the wilderness may be eaten, but this is a scavenging operation. The only confirmed cases of wolves killing human adults or children anywhere in the world have been due to animals suffering from rabies, except for two in the Gevaudan area of central France between 1764 and 1767. The two renegade wolves caused the deaths of almost a hundred people before being shot. But careful descriptions of these two particular animals suggest that they were both hybrids between wild wolves and domesticated dogs, with an unnaturally reduced fear of man.

Toward members of their own kind, wolves show a whole spectrum of reactions. Individuals that have grown up in the same pack engage in endless games of tag and prolonged bouts of wrestling with no attempt to use their strong jaws in harmful biting. One wolf may seize another by the scruff of its neck, or try to get hold of the other's jaw as each lunges at the other with mouth wide open. Some of these sessions of rough play are almost silent. Occasionally, in excitement, the competitors growl or bark softly.

A real growl, however, is a final warning to move off. A mother may produce it if playful youngsters come too close to her small pups. A few sharp barks, loud and clear, stop every wolf, no matter what it is doing, to listen alertly and discover what the danger is. A howl, particularly at night, is generally a sign that all is well but that strangers should veer off

Gray Wolf (*Canis lupus*)

beyond hearing distance. In howling, each wolf adds its own personal harmonics that the other wolves of the pack recognize.

Students of animal behavior recognize fine details in the inherited, stylized behavior of wolves and other wild dogs. The extreme in submission, for example, is that shown by a pup that is lying on its back, passively allowing its mother to cleanse it however she chooses. "Active submission" is the behavior of a pup that is approaching its mother or some other adult, begging for milk or food. A newcomer to the pack must act in these ways even after it has been provisionally accepted. And in any wolf hierarchy, low members are always ready to show submissive posture if they are challenged by any higher member.

The dominant wolf and his lifelong mate meet every challenge to their status. Particularly in the March breeding season, they become so protective that they may effectively limit the courtship activity of all inferior wolves. Only the distraction of abundant prey seems to loosen these limitations and let all of the mature wolves reproduce at close to their maximum rate. Then the natural increase in the population may be between 20 and 30 percent annually. Even this gain may be curtailed if the supply of food decreases after pregnancies begin. Until the embryos become firmly implanted in the walls of the womb, the mother is likely to absorb them as useless tissue if her hunger grows acute. Newborn pups suffer and may die if their mother must go off to hunt because her usually attentive mate cannot find a surplus to bring back to the den. Others in the pack will also be too hungry to serve as baby-tenders, for they too must range widely in search of food.

It seems lamentable that a mammal with such highly developed social behavior should have been extirpated over so wide an area of the Northern Hemisphere, while untruths about its ferocity and danger to mankind were repeated and generally believed. Now that livestock are raised less widely and draft animals have been replaced by machines with internal combustion engines, it might be possible for people to tolerate an increase in the population of these much-maligned yet fascinating creatures.

The Big Cat of Northern Forests

The silent phantom of the north woods, known in America as the Canada lynx *(Lynx canadensis)* and in Eurasia as the European lynx *(Lynx lynx)*, rarely shows itself in daylight. By night this oversized cat stalks carefully or waits on some vantage point, such as a big rock or low limb, from which it can spring upon its prey. Stealth is its trademark, and snowshoe hares ("varying hares") its favorite food.

In March, a lynx makes up for its customary silence. Males scream at one another so loudly that they can be heard over most of the six- to eight-square-mile area that constitutes their home range. The encounters that stimulate these noisy contests occur chiefly along the boundaries, which the animals patrol in the late-winter mating season more frequently than at any other time of year. Simultaneously, each male remains alert for the intense yowl of a female lynx, broadcasting her position and receptive condition with no trace of coyness. The unearthly yells of either sex make your heart beat faster and cause the delicate skin muscles to contract along your back from the waist to the scalp. Probably no mammal or bird above ground can ignore this insistent message.

Sixty days later, every female lynx that is a year or more old is likely to give birth to her kittens of the season. They are slightly larger than newborn house cats and unlike their mother in having reddish brown fur conspicuously marked with spots and stripes, whereas hers is pale gray with faint mottlings of brown. Lynx kittens open their eyes after about ten days and slowly wean themselves in two to three months, first by sucking at the fresh meat that both parents bring to the lair and then by nibbling at it. By midsummer, the kittens begin following their mother on hunting forays while the father goes off alone. Before winter, they must be able to apply the practice they have gained catching mice to capturing snowshoe hares. When the snow gets deep, each lynx is on its own.

From Alaska to Newfoundland, the welfare of lynxes is linked closely to the number of showshoe hares each lynx can catch. This fascinating dependence came to light when the eminent British ecologist, Dr. Charles Elton of Oxford University, examined detailed records kept from 1800 on by the Hudson Bay Company in Canada. The Company bought from 5,000 to 150,000 snowshoe hare pelts annually for trade; the equivalent figures for lynx pelts went from nearly zero to nearly 80,000 in a single year. These variations oscillated rhythmically every nine to ten years, reflecting the success of the hunters and trappers rather than the market. A peak in the population of snowshoe hares presaged an abundance of lynxes the following year. Whenever the hare population crashed, the number of lynxes fell off spectacularly within twelve months.

Except for man, who finds a profitable market in Europe for lynx pelts because of their long, thick, soft fur, this big cat of the northern forests has chiefly hunger to fear. With its big feet, which make tracks in snow as impressive as those of a 200-pound cougar, it is able to get about in the winter woodlands as readily as its prey. If luck matches skill, a lynx catches a hare every other night and eats everything except the skin, paws, stomach, and voluminous cecum. In a month, the predator may take fourteen hares and get from them about 41 pounds of nourishing food. Quite often the lynx returns to feed a second time on prey it has killed and only partly eaten. In a good year, a male lynx will reach full size—

three feet long with a four-inch, black-tipped tail—and weigh 40 pounds. Females remain smaller, but like their mates seem mostly muscle. With sharp teeth and claws, they could be formidable antagonists. Yet no instance of an unprovoked attack on a person is known. Most lynxes succeed in never being seen and rarely being heard except for the startling March mating calls.

The Lazy King of Beasts

On our first visit to equatorial Africa, we naively walked along a gravel road after dark for almost a half mile, from the cottage assigned to us at Murchison Falls National Park to the main visitors' encampment. We arrived in time for the evening meal, but not without misgivings along the way. Would lions or other dangerous animals be attracted or repelled by the flashlight that was our only armament against the surrounding night? Do lions give any warning sounds before they attack? What other hazards did we face, for which our earlier experiences in the tropics of the New World would be irrelevant, if not misleading?

On a later visit, we observed at length several prides of lions and many small groups of full-grown bachelors from the comparative security of a sturdy car. With us on these short safaris was the scientist who is best informed about lions in general, Dr. Charles A. W. Guggisberg of Nairobi. His personal field experiences and extensive research led in 1961 to his fine book *Simba*. Simba is the Swahili word for lion. In all the world there is only one kind: *Panthera leo*. No other wild member of the cat family is so famous and familiar, or so ensured of survival by its willingness to reproduce within the parks, wildlife reserves, and zoological gardens that man has chosen to maintain. Now that our acquaintance with wild lions has been extended, we realize how lucky we were that night in Murchison Falls National Park. Some old lion with sharp claws but poor teeth could easily have been hungry enough to learn to be a people-eater at our expense.

The list of food that lions will consume is almost endless. Zebras, which always seem to be fat, are favorites, as are warthogs. In Kenya and Tanzania, the brindled gnu (blue wildebeest) and the smaller gazelles—the Thompson's ("Tommies") and Grant's—are the most popular of antelopes. In Uganda, giraffes and ostriches are common prey. Farther south, the preferred antelope is the fleet impala. From insects to elephants, the lions get a share. And sometimes they vary this meat diet with grass, fruit, peanuts, garbage, and indigestible items of human clothing from socks to shirts.

Where the game animals are abundant, as in the national parks of

Africa, each lion may account for no more than thirteen kills each year. Consuming from 20 to 40 pounds of meat and skin from the victim, the lion satisfies its hunger for almost four weeks of indolent living. It is when 20 to 30 lions are living together in the same pride that the toll of game animals mounts up, for the family will then need more meat every day or two.

Much is heard about the strategy with which lionesses capture prey, doing all the work and then letting the lion get the choice cuts as though he needed a reward for a few well-voiced roars. Probably the sequence of the hunt is due to differences in experience among the members of the pride more than to any actual plan for action. The younger participants, including partly grown cubs, are likely to start off in search of antelopes grazing on the plains. The older, more experienced lionesses tend to remain behind and to move out in a line downwind from the grazers while the other hunters and the frolicking cubs travel at random, often on the upwind side. Their scent reaches the grazers, which then tend to drift or run toward the line of older lionesses. With luck the ambush pays off, and the pride gets a reputation for operating an efficient trap.

Most of these hunts take place before the sun gets hot in the morning, or late in the afternoon, or at night. The reason seems to be that lions and lionesses suffer from the heat; they seek out the coolest place they can find —even up in a tree—for the midday period, and scarcely move a muscle. If they are close to their latest kill, they may not even bother to chase away hyenas and vultures that have come to feast.

Below the equator, something seems to stir the lionesses as the March equinox approaches. More of them become receptive to the male then than in any other month. Since the period of gestation is 108 days long, most cubs are born in July. This degree of synchronization in the communal family benefits the cubs, because any lioness with milk will nurse any cub in the pride. The cubs are all brothers and sisters or half-brothers and half-sisters, since the dominant lion fathered every one of them. As soon as a young lion offers competition for a ready lioness, the upstart must either displace his father or be banished to join the scattered bachelors. Generally the pattern of dominance is too well established to require more than a threat to settle the dispute. Real battles between lions, and replacement of one dominant animal by another, occur more often when a male from another pride or a powerful bachelor disturbs the social establishment.

The cubs are born with extraordinarily thick coats of woolly fur, generally with dark rings around the tail and pale gray spots on head and back. These markings usually disappear by four months of age, when the youngsters begin to venture from the lair. Even at birth, the cubs are often wide-eyed. At three weeks their first small sharp teeth begin to show, although the cubs will be six months old before they are weaned. Long before that they will have found out which is their own mother, and will follow her when she lets them. Gradually they learn from her example how to get meat when they are hungry. In a national park, they are likely to discover that slow-moving automobiles offer neither food nor harm. Or, as we saw in Kruger National Park, South Africa, they may learn to walk close beside a car in its shade until the vehicle approaches a herd of unsuspecting impala. From this man-assisted ambush, a lioness can suddenly rush forward and knock down an antelope for dinner. The people in the automobile see "nature red in tooth and claw," an experience not too different from that of wealthy Romans watching lions kill in a coliseum.

In Kruger National Park, right under the Tropic of Capricorn, the mating season in March is the time when fights break out most frequently between lion and lion and between one adult lioness and another. Sometimes these bloody battles not only lead to the death of one contestant, but to acts of cannibalism by the victor—to the horror of visitors and the embarrassment of the wardens. James Stevenson-Hamilton, the former director of this park, knew some of these lionesses and lions well enough as individuals to realize that they were well fed—not desperate competitors for a limited food supply. Far fewer fights develop at other seasons, when prey might be scarce locally. Scientists and nonprofessional observers are left to wonder whether these battles to the death, followed by cannibalism, are part of an inherited program whereby the animals regulate the size of their own population before the food runs out. Or, in a park where food is unusually abundant, do these great beasts become neurotic from overcrowding and use the extra impetus of sexual drive to overcome their customary languor and commit acts of fratricide?

Modern lions go about their family life as though their privacy were complete, both in wildlife reserves and in large zoos. Every detail of their behavior is a subject for common observation. No longer need any naturalist, for lack of an opportunity to learn the truth, accept the statement by Aristotle that lion and lioness mate while facing in opposite directions, tail to tail. The lion copulates like any cat, covering his lioness-of-the-week from above and behind, often while licking lovingly at her neck. Her lack of a mane makes this easy, but may easily be a detail he ignores. In the wild, a lion may lack a mane as well if he lives in shrubby territory, for he loses his long hairs as he brushes against the bushes.

On the reserve lands, it is easy to conclude that most of a lion's life is spent in sleep and that the animal awakens only occasionally to mate or eat. At dawn and dusk he may work up enough enthusiasm for a good roar or two. Otherwise his days and nights seem uneventful. The lionesses do all the hunting and manage the cubs.

Lionesses seem to lose track of the seasons when confined in a zoo. With no antelopes or other wild animals to kill, they detect no cues from their man-made environment. They become equally receptive to the lion in any month, usually as soon as their latest cubs seem independent. Indeed, if the solicitous staff members remove the cubs as soon as they can be weaned, a lioness may produce three litters in a year—one after another. The lion is always ready to cooperate. His maleness knows no season.

The Most Voracious Mammals

Apparently there is no such thing as an old shrew. Eighteen months is a record old age for them in the wild. Sleeping briefly, never hibernating, fighting one another on most encounters, and hunting continually for food, they suffer the penalty of combining warm-bloodedness with minute size. The little body has so much surface area in proportion to bulk that it loses heat to its surroundings unless kept uncomfortably warm. The fuel for this production of heat must come from food—as much as three times the weight of the animal each 24 hours. A shrew that gorges itself in the morning can starve to death by the following midnight if forced to fast from noon on. It can also die, apparently from shock, if confined for a few minutes alone in a mesh cage.

The famous American mammalogist C. Hart Merriam once placed three lively shrews in the same small vivarium, expecting to observe them without being noticed. But two of the shrews attacked, killed, and ate the third "in a few minutes." A few hours later, one of the survivors killed and ate its companion. Having eaten close to two shrews its own size within less than eight hours, "its abdomen was much distended." In another eight hours, it was hungry all over again.

To tolerate a mate from March until September seems as much as a shrew can stand. Most of this time the female warns her mate to keep his distance. She chatters and squeaks loudly whenever he approaches. The only sound we hear from him during his courtship is an unmusical clicking. Probably his song of love is pitched too high for our ears, but not for hers. With it he tames her briefly. She lets him approach and mate for a few seconds, only to repulse him. At the peak of her receptive period she relaxes. He may copulate with her then for 20 minutes at a time, up to six times an hour, and 20 times during a single day.

All of this effort and millions of sperm cells seem needed to ensure that from three to eight helpless young the size of honeybees will be born fifteen to eighteen days later. During these weeks and until her young are weaned a month afterward, the mother may ban the male from her vicinity. He never gets to see how hairless, blind, and helpless her babies are at first, or how quickly they grow to be little shrews. Some mothers wait until one litter has gone off before copulating again. Others become receptive soon after their young are born and may produce new families at fifteen-day intervals, with six young in each.

Pregnancy does not slow a shrew in her voracious eating. She will attack a mouse several times her weight, or an earthworm, or an armored beetle. Rushing at her victim, she bites it, backs away quickly, darts in again to snap at whatever part she can reach, lets go, and tries again. Her movements are so fast that a mouse acts bewildered. Then it sags into unconsciousness, poisoned by the saliva that the shrew has injected into the wounds made by her many small teeth. Both sexes of shrews possess this chemical weapon, which is very similar to cobra venom. It upsets the nervous coordination of the prey and destroys resistance to being eaten. Each of the two large salivary glands in shrews of several kinds contains enough poison to kill a hundred mice if the gland tissue is ground up and injected into the bloodstream. The toxic saliva has little effect on predators that kill shrews, however, because they are so much bigger. Predators that decline to eat a shrew seem repelled more by the musk glands, which release their secretion along the flanks of the little animal.

The small pointed teeth that make the poisonous saliva so effective bear no resemblance to those of a mouse or any other rodent. This is one reason for recognizing shrews as members of the ancient order Insectivora, whose early representatives were contemporaries of the giant dinosaurs. The huge reptiles vanished, but the minute mammals lived on. At least one kind, the water shrew *(Sorex palustris)* of Canada and several mountain states, is so quick and light that it can scamper across the surface of a quiet pond without sinking in. Bristles on its small feet trap air and provide buoyancy. They also help the shrew swim if, on its overwater journey, a wind creates waves and the mammal gets wet. One of the first live shrews we met was crossing a lake a half mile wide. In proportion to its three-inch length, this trip would correspond to a twelve-mile swim for a six-foot man. Although we matched its progress in our skiff for the last 100 feet, it never wavered from its course. Once ashore, it shook itself dry, groomed its fur, and ran off into the forest undergrowth. Probably its rate of heartbeat had already slowed down to the normal 800 per minute!

The March Madness of Pocket Gophers

Pocket gophers live underground in the plains and semiarid lands of western North and Central America. Although the females exhibit a most shrewish disposition, both sexes are larger and more powerful than any shrew. They are members of the little family Geomyidae (literally "earth mice"), and grow to be eight to ten inches long, weighing only a pound or two. Their compact vigor is evident in the conical snout, small eyes and thick neck, sturdy forelegs with long, strong claws, and a stout body that ends in a short, naked tail. Under the jawbones are two external openings into the fur-lined pockets for which the animals are named. The pockets extend backward to the shoulder region, and are often packed with vegetable trophies that the gopher is carrying to its underground storage rooms.

These American rodents ordinarily stay in their tunnels, moving through them at surprising speed. Neither a predator nor a person has much chance to see what goes on in the extensive underground apartment. Then, in March, the ardent males abandon their customary caution and their underground safety. They emerge at night, and sometimes by day (particularly in cloudy weather), to scamper about over the surface. The mating season begins in the South and comes later in the Northwest.

We feel sorry for the males, for they risk their all and so seldom find a mate in a cooperative mood. So far as has been discovered, each female remains hidden, not even unplugging the door to her passageways to help odors attract a male during her brief time of sexual need. She makes no audible or visible signal. Only the reckless persistence of the males accounts for the pregnancies that begin in March and last for 28 to 30 days. In regions where predators are numerous, many a female pocket gopher receives no visitor of the opposite sex. Her annual reproductive season passes uneventfully and unproductively.

Recapture of individuals made distinctive with numbered metal ear tags shows that the male returns, if he can, to his own system of tunnels. A pregnant female continues for a while with the ordinary chores that are essential for either sex. She remains alone in her burrow, extending it to reach food underground and defending her domain ferociously from any intruder, including those of her own kind. These animals travel hundreds of feet daily in passageways of their own construction. Usually the maze includes two distinct levels. The most extensive tunnels lie just a few inches below the surface of the ground, with a roof so low that the owner must move in a perpetual crouch. At intervals the pocket gopher bypasses temporary storage rooms, or steps across vertical sump holes that serve as traps for any rainwater that gains entrance and flows along. These are the feeding tunnels, with many side branches to let the burrower reach edible roots. The favorites are those of weeds, which have special adaptations for growth during droughts and cold periods when grasses are likely to be dormant or dead. By selecting the concealed parts of weeds and shrubs, pocket gophers actually bias the ecology in favor of grasses and perpetuate the grasslands.

Deeper passageways, five feet or more below the surface, extend to chambers beyond the spread of frost. Here an adult gopher is likely to have at least a dozen rooms, a few of which are used for storage of food, while one is the current bedroom and another a toilet chamber. The animal is careful never to wet or foul its bedding, but it never tidies up the leftovers from frequent snacks that it has taken to bed. Eventually the spoiling food attracts so many insects, mites, and spiders that the gopher abandons the swarming bedroom and digs a fresh one. The animal is more fastidious about the toilet; after this room has been used for a week or two, it is closed permanently with vegetable debris and earth. Another room, whether a former storage chamber or a new space, is then selected for this purpose.

As the day for giving birth approaches, the pregnant female digs a new room in her subterranean apartment. With help from her prominent front teeth and the curved claws on each forefoot, she loosens the earth until it can be pushed past her body. Working rapidly, she accumulates so much behind her that she must turn end for end and act like a bulldozer, forcing the debris along her passageway. Some of it goes into unused chambers. But much of the earth gets shifted progressively all the way to the front door. Since this is plugged with earth, it must be opened cautiously. Sunlight or the odor of a predator nearby induces the pocket gopher to ram all of the available material into the entranceway. Darkness and the smells of adjacent vegetation only encourage the animal to disperse the transported earth in a fanlike pattern as much as four feet long, sometimes to a depth of six inches to a foot. However, the last load of earth is always saved to make a new tight plug for the doorway. It is a thick plug in badger country, for a hungry badger can dig almost as fast as a pocket gopher and will make a powerful attempt to reach its prey.

The new nursery and some of the storerooms are well stocked with fresh, fine rootlets and, if possible, some green grass or other fibrous material collected outdoors at night. These provide bedding for the fat newborn gopher, whose naked pink skin falls loosely into folds and wrinkles. The litter may consist of just one addition to the population, or twins, or as many as nine, depending partly on the age of the mother and partly on the species of pocket gopher. At first in utter darkness, the young seem especially dependent upon the mother, for their eyes and ears are sealed shut. She begins to supplement her milk with vegetable foods when they are three weeks old, although their eyes will remain closed for another two weeks. Yet at six to eight weeks of age they lose their welcome in her tunnel system and are thrust out—usually at night—and discouraged from ever returning. If they cannot find a vacant burrow close by, they must risk further travel or dig in for themselves.

The first 24 hours away from their birthplace hold the greatest dangers for pocket gophers. Owls by night and hawks by day are ready to pounce on the wanderers. Foxes and coyotes watch for them. So do skunks and house cats. Badgers can run them down or dig them out. Weasels and several kinds of snakes enter any tunnel system that is not yet tightly plugged. Nor can the growing antagonism of one pocket gopher for another be forgotten. Already the young animals are showing a mutual intolerance that becomes a mania, with each individual ready to fight to the death any other that enters its burrow or comes close when both are on the surface of the ground. Young gophers take extra chances with predators because, for a while, the taste of fresh vegetation above ground appeals to them much more than that of roots. Gophers that survive despite foraging outdoors close to the burrow mouth generally develop a pronounced preference for the foods they can find by tunneling below the surface.

These food habits must be synchronized with the seasons, for pocket gophers choose different roots at different times of year. The temperature in the tunnels closer to the surface certainly changes too, and with it the need for heavy winter fur. March is not only the time for the males to go mate-hunting. It is also the beginning of the annual molt, which starts around the tail of the animal and progresses slowly forward until, by early fall, it reaches the nose. Through the summer a belt line marks the body all around, where old fur has been shed and new hairs have been exposed.

We wonder how many millenia ago the first of the pocket gophers settled into this way of life. About 30 different kinds now mine for food and shelter from Panama to Calgary, Alberta, and around the Gulf of Mexico to Florida. They suit a peculiar niche in the living world, one that can be found from sea level to mountain slopes more than 13,000 feet high, well above the timberline. The vegetation in these many habitats shows no uniformity, for the plants are those of arid deserts, semiarid grasslands, sandy ridges in the Gulf States, valleys, and open forests. The unifying themes seem to be darkness underground and chronic solitude, punctuated by short deep sleeps but no real hibernation. It is a life of all work and no play—not even when March and mating season come around.

The Rabbity Tribe

Long ears and long hind legs distinguish these familiar vegetarians (family Leporidae), in which females (does) normally grow larger than males (bucks). Hares are more proficient than rabbits in running, rather than hiding, from danger. Hares make no maternity nests, and bear young that are open-eyed at birth, able to move and fend for themselves in hours. Rabbit young are born naked, eyes sealed shut, in nests three to four inches deep and eight inches across, blanketed by fur that the mother pulls from her breast and belly as she exposes her nipples for nursing. Rabbits and hares equally develop by adulthood in the males a scrotum containing the testes located in front of the penis, rather than behind it as in all other mammals except marsupials.

The most famous member of the rabbity tribe is the European hare

(Lepus capensis)—the "March Hare" of *Alice in Wonderland.* Slightly more than two feet long, with a three-inch tail concealed in a powderpuff of long hairs that are black above and white below, it is so timid an animal when adult that it often seems unbelievably stupid. Early in each day it returns to a saucer-shaped depression, called a form, that it has scraped in some dry sheltered place, or else makes a new one. There the hare remains motionless, its long ears laid smoothly on its brown back. Its protruding eyes provide panoramic vision from the zenith to the horizon in all directions. Only its mobile nose twitches, sometimes swaying the long projecting whiskers, as the hare tests every breeze for an odor that could mean danger.

In early March the male hares become more daring. Both by night and day they risk detection by exploring thoroughly the few acres of territory each regards as home. If another male is found, the defender rushes forward and rises on his long back legs while striking at the other with curled forepaws. The intruder may reciprocate, in a furious boxing match. Soon one hare—usually the stranger—tires of the contest and runs off, leaving the area and any female in it to the victor.

Female hares are less limited to particular fields or woodland borders. They tend to take short steps or jumps if a male presents himself before they are ready. The courting male persists, trying to caress the female with lips and forepaws, twitching his tail and often urinating in his excitement. When she does allow him to mate with her, she cooperates for a few seconds, then throws him off and pretends to be hungry. After a minute or two, she stops eating and lets him get into position again. The mating position is not tail-to-tail as Aristotle concluded. Although the penis of an unexcited hare or rabbit is directed backward, and can be observed in this orientation in a dissected animal, the organ curves forward during copulation.

The growth of the young hares, called leverets, takes five weeks or more inside the mother. A few days before they are born, she is likely to accept the male's courtship again to ensure that the next eggs from her ovaries will be fertilized and ready to take their place in her womb as soon as it is vacated. Her first brood of the season is likely to consist of no more than two leverets; her second, four; her third, three; and a fourth brood (if the weather holds), one or two.

A newborn hare defends itself as best it can from its first day on. Until it is weaned at the age of two or three weeks, it will hop forward like a big furry toad, growling in an astonishingly low voice, ready to bite at anything that disturbs it. After the leveret has grown bigger, it abandons this reaction and instead takes to its heels, relying on its long legs and tremendous leaping ability to speed it away from danger. By eight months of age it will probably be sexually mature and pursuing potential mates, or be pursued as a likely parent of more March hares.

Something in a hare's constitution seems to prevent the population from growing as rapidly as might be expected. Pregnant females can be found in any month from January through September. Yet early and late in the season, the unborn embryos tend to disappear, being absorbed rather than aborted. This is why a mating in March appears to be most fruitful. Of the seven to eight leverets born to the average female in a season, almost half open their eyes in April and early May. Over much of the hare's range in central and southern Europe, the Middle East, and Africa as far south as the Sudan, the new arrivals begin nibbling on plant foods while fresh growth from springtime is still available.

These traditional ways persist among the European hares that have been introduced into North America. Over an area from Massachusetts to Minnesota, chiefly north of the Great Lakes, they are often mistaken for jackrabbits, which are not found quite so far east. The range of the introduced hares overlaps instead with their native kin, the snowshoe (varying) hare.

The snowshoes are well named, for their long toes spread widely and almost seem webbed in the winter season because the coarse hair on the soles grows longer than in summer; it prevents snow from coming up between the toes and keeps the hare atop the snow. For much of the year, including most of the breeding season, this adaptation for quick progress in a snowy world holds great importance for the hare. Its habitat is the forest as far north as trees grow, south in the western mountains into California and New Mexico, south in the Appalachians to Tennessee, and elsewhere chiefly as far south as the states that border Canada. Each individual, however, rarely strays more than a sixth of a mile from the place where it was born. In summer, when its fur has changed from snow-white to brown, it travels even less widely.

Within its territory, a snowshoe hare learns every twist and turn. If pursued, it circles to stay close to its familiar hiding places. It spends most of the day in one of its forms, or under a log or other shelter. Night is its preferred time to be abroad, although moonlight seems quite acceptable. Between sundown and sunup the animal explores for food and during much of the year ignores others of its kind. From August until late January, a male has no interest in the members of the opposite sex. Few females are ready for mating before mid-March; early April is soon enough for many of them. Yet before the end of August, some females will have borne three litters totaling nine young.

The persistent mating habits of snowshoe hares give little hint that their fortunes rise and fall spectacularly in cycles lasting eight or nine years. Each surviving hare attains sexual maturity during its first winter, before reaching one year of age, and takes every opportunity for promiscuous mating with members of the opposite sex. A male that finds several receptive females on the same acre of woodland proceeds to copulate with every one of them in turn. Sometimes he loses count, and copulates a second or third time with a few of them as he makes his rounds. A receptive female will accept a half dozen suitors in quick succession. Under these wild conditions, male parentage is impossible to trace but pregnancies seem assured. An exploding population might be predicted, but hardly a declining one.

For some reason that has still to be discovered, the number of young in a litter and the number of litters in a year vary greatly. In this feature lies a far better reason for calling the snowshoe hare a "varying" hare than the change in the color of its coat between winter and summer. Before the population reaches its peak size, the reproductive output of each mature female falls off dramatically. No longer does the birth rate equal the death rate, let alone exceed it. The number of snowshoe hares plunges until an area that supported 150 of them may have only 10 remaining. Attempts to attribute the abrupt decline to exhaustion of food supply or hiding places, or to diseases and predators, have been unconvincing. Perhaps the female responds to the stress of moderate crowding by reabsorbing the embryos that are developing in her womb. The unsuccessful ending of pregnancies that have started is a form of birth control that results in negative population growth—not just zero growth. Only when the population has shrunk drastically do the female hares attain their full reproductive potential once again. The numbers of these fleet-footed denizens of the northern woodlands then begin to increase, but perhaps not before their depletion has brought near disaster to long-lived predators such as the Canadian lynx.

South of Lake Superior and east of the Rocky Mountains, the place of hares in nature is occupied mostly by the cottontail *(Sylvilagus floridanus* and *Sylvilagus transitionalis),* the wild "bunny rabbit" familiar to many youngsters. Western relatives inhabit mountains and deserts in the Southwest. Everywhere the cottontail responds to the longer, sunnier days of March by increasing its sexual activity; long periods of cloudy weather depress its reproduction. Winter cools both the ardor of the males and the ability of the females to become pregnant. Consequently, the population of these three-pound hopping herbivores ordinarily sinks in late February to its smallest for the year.

If we sit quietly in late afternoon or early morning beside a field or neglected lawn where the spring sun is encouraging the plants to produce fresh growth, the cottontails are likely to appear. They enliven the scene early in March through the southern part of their range, and later in that month farther north. First one individual will hop from concealment and creep into the open in slow motion, feeding continuously. We cannot tell whether it is a male or a female. But if another cottontail emerges from the undergrowth, the interactions of the two give away this private detail

Cottontail rabbit (*Sylvilagus Floridanus*)

of their lives. A male will begin following a female, sniffing at her tracks, and coming up close behind her as though trying to learn her sexual condition by scent. Generally she whirls to face him, and the two stand nose to nose for several seconds. She may leap right over him, as though playing leap-frog. Usually he turns and tries to follow again. She spins around and jumps over his back so quickly that he cannot touch her. Yet she often leaves her mark by urinating on him as she jumps. Time after time the two may repeat this peculiar ceremony until, his ardor dampened and his fur wet, the male departs to clean himself and find a more cooperative mate. If he does not, she may rear up on her back legs and strike at him with her forepaws until he gets her message.

A cooperative cottontail is one that will allow him to mate with her for as little as ten seconds or as long as seven minutes. In March this readiness in the female may match the concentration of several hormones in her bloodstream—those that have to do with shedding her winter coat and growing a lightweight covering of summer fur. Toward the end of her pregnancy, which lasts four weeks, she will carefully comb out many of the long hairs from all parts of her body except her abdomen and use the soft fibers to shape a nest among the undergrowth. In it she will bear four to seven young, each weighing a mere three-quarters of an ounce. For a few hours her attention is focused on giving birth and cleaning these helpless, hairless, pink, wrinkled additions to the cottontail population. She nurses them by standing above the nest while they reach up to the nipples on her underside. Then she carefully closes the roof over them and hops off to meet the male—usually the one that has been following her persistently for the past three days. Within minutes she has mated again.

By comparison with most mammals that we know, the female cottontail seems strange in releasing eggs from her ovaries only after she has already received from the attendant male the sperm with which fertilization can take place. Perhaps, through some quirk of chemistry, she can nourish the sperm cells while they wait for her eggs to arrive for more days than she can keep her eggs alive while waiting for a male to supply the sperm. But as the days shorten in autumn, her willingness to produce another litter drops to zero. By then her hormones are calling for the growth of her winter coat. She can no longer spare the long hairs with which to insulate a further batch of kittens against October's chill.

CHAPTER TWO

April is a Quirky Month

MORE THAN A score of springtimes have swept poleward through the North Temperate Zone since we first began keeping a record of the rhythmic changes in our particular valley. There the pace of life quickens in April, inviting us to enjoy a brief repetition of the springs from previous years. We must capture this fleeting sample or wait three seasons to meet it again. Before the sun becomes too hot and the moisture scanty, every wild neighbor hurries to reproduce, to feed its young, and to get the most out of the lengthening days.

The weather changes abruptly from a reminder of February to a foretaste of June. On rivers and ponds that have long been roofed with ice, margins and avenues of open water appear. Buds open on one kind of vegetation after another, making us favor the contention of some philologists that April, from the Latin *aprilis*, came originally from the verb *aperire*, to open. But it is a discontinuous, quirky opening, never a steady progression. No one can be sure what will happen next.

Largest of the "Picket Pins"

Alertness for detecting change seems exemplified in the marmots, groundhogs, prairie dogs and chipmunks, all of which sit upright at their nest holes like so many picket pins. First to do so in April is usually the groundhog or woodchuck *(Marmota monax)*, lean after a winter's sleep, ready to eat hungrily under a sun that casts far more shadow in our part of the world than on Groundhog Day (February 2). After a good meal or two, these familiar animals are eager to meet a mate. At first they have plenty of time to fight, but little for courtship. When one male meets another, they battle viciously, growling and grinding their cheek teeth. They snap at an opponent with sharp, chisel-like incisors, and sometimes bite off a paw or part of the rival's tail. Whichever individual gets the worse of the contest runs off limping.

A male shows none of this territorial belligerence with a female of his kind. He examines her with his nose, and reacts promptly if he finds her ready. Generally she accepts the first male to come along. A minute after they meet, she may be pregnant.

Following the first brief breeding season of the year, woodchucks show more individuality. Many go off to almost solitary lives. From the sandy doorstep of the burrow, they watch for neighbors and any approaching fox or hawk. At the first sign of danger they whistle sharply and vanish underground, only to emerge shortly afterward for a cautious study of the situation. Nearly a third of these solitary groundhogs will emerge to eat if they can see a neighbor similarly engaged. The rest turn back into the den if another woodchuck is in view.

A good many males seem to move in with the first receptive female they find, and share her burrow for the rest of the summer. Other males visit one female after another before settling down. By the time the females give birth, 28 days after mating, all of the groundhogs have had an opportunity to improve their living quarters. The best burrows are under trees, where a forking root prevents any fox or badger from easily enlarging the doorway enough to enter. One groundhog that we know quite well is even more secure because the entrance to her apartment in our hillside is between two huge granite boulders, left one atop another by a retreating Ice Age glacier. Her front door is a hole so narrow that in autumn she can barely squeeze through it herself.

The young of the year will be born in May or early June. When they outgrow their initial nakedness, blindness, and general helplessness, they begin moving about under their own power. They never become housebroken until after they have been weaned, and we can imagine their mother being impatient for them to grow up enough to go outside the burrow. She herself is quite fastidious, replacing the soiled grasses under her litter with fresh ones several times a day. After about five weeks of this routine, she brings her offspring outdoors and encourages them to save her labor. Now they will take solid food instead of milk. Often they dig a den of their own, usually near by, but learn about diet and dangers by continuing to browse close to their parent. More often, the maternal quarters are enlarged to shelter the whole family, and the growing young-sters remain even closer to their mother. A good many burrows are big enough that the groundhogs can also tolerate a rabbit or two underground with them.

Each groundhog gets fat as the summer progresses. An adult that weighed five pounds in April will weigh eight to ten pounds by September. Animals less than a year old are only slightly smaller. All of them have ceased to concern themselves with territorial disputes. They are too busy building adipose tissue inside their skins as a food-store to last them through their winter sleep. To us, this lessening of competition with other groundhogs makes sense. After all, since woodchucks sleep at night and are abroad mostly at daybreak and toward sunset, any fighting they do is in good light—enough to make them visible to sharp-eyed predators. Before a combatant, whether winner or loser, could run to the safety of a burrow, it might lose its life. Groundhogs ordinarily have no surplus population to risk in this way. The natural increase in their numbers allows only for a doubling of the population each year. After the April battles and the occasional loss to a beast or bird of prey, parasitic infesta-tions and diseases usually prevent any significant multiplication of groundhogs.

We have often wondered what message from the fall season reaches all the local groundhogs in the same week, and sends them to the depths of their respective dens to sleep until spring. A scientist at North Carolina State University, David E. Davis, decided to find out. He experimented with all the woodchucks he could reach—thin ones, fat ones, and middle-sized ones. Every one of them reacted in the same way. In autumn, the animal retired for at least four days, and often five, if for a whole week it could not find enough suitable food to keep its stomach full, and if the temperature outside stayed in the low 40s Fahrenheit. A second such experience seemed conclusive in initiating true hibernation. Yet, even if for eight consecutive months, the groundhog finds no dietary or thermal encouragement outdoors, it still interrupts its hibernation every ten to twelve days to check on the possible arrival of spring and fresh food.

Other Squirrels That Sleep Underground

To a mammalogist, a woodchuck is a kind of marmot, and marmots are a type of large ground squirrel, to be contrasted with tree squirrels and flying squirrels. When threatened, a ground squirrel goes underground, whereas a tree squirrel goes up a tree. In North America, the other ground squirrels include the golden-mantled and the thirteen-lined ground squir-rels (both *Spermophilus* species), several kinds of chipmunks (genera *Tamias* and *Eutamias*), and the prairie dogs *(Cynomys)*. Of these, a few (particularly the golden-mantled ground squirrel and the black-tailed prairie dog) resemble groundhogs in beginning their reproductive year when April arrives. Apparently it is a common family trait; in Eurasia, the ground squirrels known as sousliks *(Citellus* species) often behave in a similar fashion.

The golden-mantled ground squirrel *(Spermophilus lateralis)* makes friends easily in our western national parks. Somewhat larger than a chipmunk and striped along the flanks, although with no dark stripe through the eye, it turns up at picnic sites and learns to scramble for peanuts that are tossed in its direction. Often we have watched these engaging squirrels compete with the large western jay known as Clark's nutcracker, which also enjoys peanuts. The bird will take to its wings if rushed at by a hungry squirrel, but can travel airborne to the next peanut faster than the little rodent can on its short legs. Either creature can dispose of a supply of peanuts at high speed. The squirrel rushes about like an animated vacuum cleaner, storing them temporarily in its capacious cheek pouches until it seems to have the mumps. The nutcracker picks up the kernels one at a time and slides them into an equally adequate crop. Both beggars then leave. In a few minutes the golden-mantled ground squirrel is back again, ready to take another load to its underground storage center. The bird must wait until digestion has caught up with its big meal.

Ground squirrels store dry foods, including desiccated mushrooms, in an underground retreat. Meat is eaten wherever it is found. It may be a grasshopper, a caterpillar, a big fly, or a beetle. If the ground squirrel discovers a nest of mountain bluebirds during explorations up a tree, it will dine on the nestlings.

Each ground squirrel has its own hideaway, although most seem to follow the same basic plan in excavating it. From the front door, which may be three inches across, the tunnel narrows to about two inches as it slopes downward at nearly 45 degrees, then levels off. At eight to fifteen feet distance, the passageway rises again to a hidden rear exit, which is reserved for emergency use. In addition to the side tunnels used for storage, the squirrel digs a circular den about midway along the main passages. Generally this chamber is eight inches wide, with a ceiling at least four or five inches above the floor. There the squirrel prepares a crude nest of plant fibers. It is thin in summer but thick enough in winter to give insulation above and below, like a mattress plus a blanket. When the bedding wears out, it is transferred into a storage chamber and replaced with fresh material. This habit probably lessens the likelihood that a badger or a bear will find odorous evidence of a squirrel's presence and try to dig it out for dinner.

While begging or foraging for weed seeds among the undergrowth, each golden-mantled ground squirrel remains silent, as far as our ears can tell us. But when a golden-mantled ground squirrel rushes home to seek sanctuary, after it pauses on its doorstep for a final backward look, it whistles a nervous *tsip!* and disappears. Later, when it looks out again, it may begin an irregular ticking sound if it can still notice a change in its vicinity, such as a human family picnicking near by. This is entirely different from the buzzing and chirping noises that the squirrel produces to signal antagonism to trespassers of its own kind. Both styles of message are generally accompanied by a conspicuous flicking of the tail.

Some of these squirrels become excited enough to attack a neighbor who ventures too close. The battle may be violent, but is usually brief. More often, we have seen a golden-mantled ground squirrel turn tail and scamper for its burrow if challenged by the much smaller western chipmunk. The first time we saw this happen—a four-and-a-half-inch animal putting a ten-inch one to flight—we concluded that some danger unseen by us must have frightened away the larger squirrel. Not until we realized that the smaller one won almost every contest did the outcome impress us as paradoxical. It cannot be explained on the basis of competition for food, since except when offered a peanut their diets differ completely. Both rely upon plant foods for most of their energy and will feast on the common dandelion if it is available where they live. However, the golden-mantled ground squirrel chooses the stems and rarely the flower heads or seeds, whereas the little chipmunk strongly prefers the flowers and seeds.

Only in the springtime, a few days after emerging from hibernation, does a golden-mantled ground squirrel hold its tongue when it sees another of its kind. Then the males creep quietly toward any neighbor and investigate. Females pretend not to notice the advancing males. Perhaps each male is singing a love song in the range of ultrasonic pitches a human ear cannot detect. Probably these courtship antics have accompanying odor messages too. These squirrels possess special small glands just behind their shoulders that secrete an oily liquid with a musky smell. Like other ground squirrels and marmots and some less closely related rodents, they also have anal glands that give off a distinctive aroma. We wish we could tell what chemical communications go back and forth between potential mates. So alert are these animals, even in mating season, that we can never get close enough to convince ourselves.

Location affects the timing of courtship, just as it does plant growth. On south-facing slopes in the Rockies and Cascades, the golden-mantled ground squirrels begin mating earlier than on north-facing slopes. They are later in northern British Columbia and Alberta than in southern California and west central New Mexico. Their amorousness follows spring from the valley toward the peak. If an average date could be estimated, it would come in April. Some of this variation is cancelled out in the timing of subsequent births, probably because of a delay between fertilization of the eggs and implantation of the embryos in the walls of the womb. After implantation, less than four weeks are needed before the young are ready to be born. This event comes late, usually toward the end of June. July is half gone before the babies have their fur and are alert and self-reliant enough to go outside the burrow. By then they seem to be more than a quarter of their final weight. A week or two later they are independent.

Golden-mantled ground squirrels retire underground before 6 P.M., presumably to bed, and do not appear again before 10 A.M. The squirrel sleeps all winter, from September until April if it is three years old or more. Younger squirrels of this kind stay active for about a month longer in the fall, enlarging their burrows and laying in a bigger store of dry food. All ground squirrels sleep solidly. During hibernation they can be handled in bright light without awakening. For at least six months of every year and twelve hours each day, the animal shuts itself away from social life. A quieter existence would be hard to imagine.

By comparison, the black-tailed prairie dog *(Cynomys ludovicianus)* lives a hectic social life. This lowland cousin of the golden-mantled ground squirrel is native to the short-grass plains between south central Canada and northern Mexico. It is a stockier counterpart of the white-tailed prairie dogs of the Rocky Mountain region. On a black-tailed prairie dog, the black-tipped tail is fully a fourth the length of the animal's body, whereas that of the white-tailed species is white-tipped and less than a fifth as long as the body of the slimmer rodent. Formerly, black-tail communities stretched for hundreds of miles. Today a prairie dog town is almost as rare on the Great Plains as a herd of bison. The surviving animals of this kind live mainly in a few national parks, where no rancher can poison them to preserve forage that a starving cow might eat.

Soon after sunrise, the black-tails climb cautiously from the single large opening that serves a group of them as perhaps the only doorway. With black eyes set high in their heads they peer over the surrounding rim they have constructed. This gives them a slight elevation above the flat land on all sides, and also protects the burrow from flooding during any sudden downpour. If no danger is visible in any direction, they venture out on the rim itself, crouching low and examining all of the territory and sky within view. Any dark fleck circling overhead could be an eagle or a big hawk, such as a red-tailed or a rough-legged hawk. Any new mound on the grassland might be a crouching coyote, a fox, a bobcat, or a badger. It need not be high to show, for the black-tails keep every bush and grass plant trimmed to no more than six inches high within 75 to 100 feet of the doorways. Under all those searching eyes, an intruding predator almost surely will be discovered. The warning spreads like wildfire, with every prairie dog barking a sharp protest or a series of protests before retiring to the depths. The retreat may be gradual, leaving a few sentinels to continue the sporadic warning. Or the prairie dogs may tumble head-long down their passageway and sort themselves out at a safer level.

The all-clear signal is equally a group responsibility, although we have no doubt that among all the black-tails checking for danger, some few older and bigger individuals are the leaders. Yet even from a movie record, it is not easy to decide which prairie dog is the first to rise abruptly from a sitting position, throw its head back and utter a whistling cry, and at the same time strike at the air in front of its chest with both front feet at once. Almost simultaneously, every prairie dog in town makes the same gesture and call. Often the animals have no sooner dropped to touch the ground with their forefeet than they go through a second round, as though to make sure that every member of the community receives the message. Then they move off into the feeding territory between one burrow opening and the next.

In the areas where prairie dogs still have a free choice among native foods, they select a diet that is about 45 percent grasses, 25 percent plants of the goosefoot family, 28 percent vegetation of other kinds, and 2 percent insects. Leaves, stems, seeds, and fruits are all acceptable. So are the roots of woody shrubs, such as sagebrush. Where black-tails maintain a thriving colony, sagebrush is scarce. On many parts of former prairie-dog territory, where the rodents have been systematically destroyed for the benefit of cattle, sagebrush now maintains an almost solid stand—shading out the plants that cattle would accept as food.

After a community of black-tails has become accustomed to our parked automobile, we can watch through field glasses while the animals are busy eating. They seem to take turns at sitting up extra tall and looking around as sentinels. Often they use their forepaws like hands to manipulate grass heads and the tops of other plants while nibbling out the tastiest parts. Occasionally the approach of a prairie dog frightens a grasshopper into leaping, and the rodent takes after the insect, perhaps catching and eating it before it can get into position for another jump. Caterpillars and beetles need less pursuit. Other, smaller insects that have been identified among the stomach contents of prairie dogs seem to have been eaten accidentally, along with the vegetation.

So many foods appeal to a prairie dog that hundreds of these animals can stay plump and active on a range where cattle are lean and hungry. Nor do the rodents suffer obviously from thirst, for they benefit from dew in the early morning and by hiding underground from the dry heat of midday. Unable to understand this difference in adaptive fitness for a semiarid environment, ranchers have often claimed that each prairie-dog town had its own deep well, dug by the inhabitants. Many a professional well-driller has gone bankrupt trying to find water under a prairie-dog settlement, sure that an underground stream must be there somewhere to support so many animals.

The burrow itself commonly goes no more than fifteen feet below the surface. At three feet down or slightly more, the plunging entranceway is cut out at one side to provide a shelf or waiting room. There a prairie dog or several can stand listening with small sharp ears for news from the outside world. At this level, the main passage may zigzag slightly before continuing downward. This is particularly common where, below the grassroots level, the soil has a limy hardpan that limits the downward percolation of water in rainy weather. At greater depths the tunnel system divides into horizontal halls with rooms of various sizes. Some are bedrooms, with clean mats of plant fibers on the floor. Another may be a toilet, for most prairie-dog families are fastidious. None are storage rooms, however, for these rodents are unlike most ground squirrels in never hoarding. Instead, they are more like groundhogs, building up a fat reserve under the skin and relying on it to tide them over all periods of food scarcity.

Often one or more of the side halls ends in an upward-slanting portion that once opened to the surface. It served temporarily as a route through which earth from the burrow could be removed. Now it is firmly plugged again with soil. All of the prairie dogs in a single burrow system use the conspicuous front door for their goings and comings.

Strangers are unwelcome underground, although visiting back and forth on the surface is common enough. Even in late March, when the mature males begin going from one doorway to another searching for

Black-tailed prairie dog (*Cynomys ludovicianus*)

mates, they chatter loudly and approach with care. Fully half of the yearling females, and almost all of the older ones, become guardedly hospitable in April. They receive the males on the front porch; copulation takes place in the open, close to the female's tunnel. Unless a predator threatens at that moment, sending every prairie dog below, the male must move on. Often he has a long way to go and will mate several times with a series of females before arriving home. The adult males tend to congregate and build their tunnel systems in one portion of the community feeding grounds.

Each female has her own apartment, and it is there that she gives birth to her litter of pups. If she is a yearling, she may have no more than two or three. Older mothers usually bear twice as many. In the unrelieved blackness of the inner chamber that serves as nursery, she sits up to let several get milk at once. Until after the middle of June, however, we never have a chance to see the youngsters. Not until they are about seven weeks old will they begin following their mother to the surface. Generally they move about in a confused heap, each youngster falling over the others in an attempt to stay close while examining the strange, bright world. When they rush back to the front door of their burrow, which happens frequently with no obvious cause, they tumble in and then emerge again immediately to peer around—unless a stern-sounding bark from their mother sends them below at once.

As the young prairie dogs are weaned, their mother leaves them progressively more on their own and finally deserts them. Their inheritance is the burrow system, while she sets off to find an empty one or to dig another for herself. If they survive, they too will disperse a few weeks later. A surprising number will discover vacant apartments waiting for them and move in, ignoring if necessary any dead relative they may find below ground. Apparently it is only incidentally that a prairie dog will fill with earth or discarded bedding the chamber containing a corpse. Entombing the deceased is not an inherited ritual among these animals, as was formerly believed.

Each prairie-dog town is a social entity, defended against outsiders of the same species by citizens of both sexes and all ages except the young of the year that have not yet dispersed from their mother's burrow. With these same exceptions, every prairie dog has a personal chamber in the burrow, which it enlarges from time to time. We see only the last stages in this operation, when the rodent backs out of the communal doorway, kicking between its legs the particles of earth that it has loosened deep underground by scratching with the claws on its forepaws. The earth may be used to repair the dike around the doorway, or be spread still farther afield.

Until European people arrived on the Great Plains with domesticated animals and a determination to protect them from every hazard, the prairie dogs participated in an amazing balance of nature, a balance which simultaneously benefitted the great herds of bison, antelope and elk, birds and beasts of prey, nomadic tribes of Plains Indians, and the vegetation upon which they all ultimately depended. Now most of this ecosystem is gone.

Any rancher realizes that his cattle and sheep may stumble into a burrow opening, break a leg, and then starve from an inability to walk around. At least some of the food that prairie dogs eat could be used profitably by domestic animals. In both ways, the rodents reduce the efficiency with which the land can be employed in producing livestock for human needs. Generally overlooked, however, is the degree to which prairie dogs keep the grasslands free of woody shrubs, let rain and snow-melt filter into the soil, slowly overturn the soil itself, and actively prey upon grasshoppers—all services a farmer appreciates.

People whose interest in prairie land extends beyond taking away everything it can yield for profit are more likely to regard prairie dogs as fascinating neighbors. Could any human population sustain itself in such balance with its world, with so little outright aggression and so much social intercourse? Prairie dogs demonstrate in their inherited way of life a versatility in coping with annual changes in the supply of food, with the daily routines of climate, and with all the natural hazards—including prairie wolves, grizzly bears, and Plains Indians.

Ordinarily we expect a mammal with versatile ways to fare better in an environment altered by man than in one with stringent requirements relating to food and habitat. Yet such are the vagaries of interaction between wild animals and civilized mankind that the versatile prairie dogs have succumbed under campaigns to safeguard the plains for livestock, while the more limited western tree squirrels have survived despite efforts to cut the forest trees for human use.

Tree Squirrels with Tasseled Ears

Some of the western tree squirrels are affectionately known as the "tassel-ears" *(Sciurus aberti)*. They limit themselves to the forests of ponderosa (yellow) pine from northern Colorado down the mountains to the Mexican state of Durango. Throughout these highlands, stretching eleven hundred miles north and south, they eke out a tolerable living during the snowy winters by cutting short lengths of pine branches and eating the thin sheath of nutritious cambium immediately below the bark.

Ponderosa pines seldom offer tree holes in which a two-pound, twenty-inch squirrel can curl up snugly to escape the cold wind. The tassel-ears rely instead on windproof shelters they build for themselves high above the ground. Each one seems shaggy, but is woven together skillfully. The outer wall is of pine branch tips with the evergreen needles still attached. Within this is a soft lining of dry grasses and frayed bark. Often the size of a bushel basket or more, the nest shields the squirrel from the worst weather. It is occupied not only during the winter season, but also when rains continue hour after hour later in the year.

The tassel-ears explore abroad on all fine days. They hop about on snowfree ground or search among the tree tops. They dig among the loose pine duff for mushrooms and lupine seeds, as well as for fruits of many other herbs and shrubs. In season, they venture beyond the pine forest among the western oaks to dine on acorns. Sometimes the squirrels clean out the cached acorns that the California woodpeckers in these mountains have providently hammered into crevices of tree bark and into cavities of their own creation in telephone poles.

By mating in early April and giving birth 40 days later, the tassel-ears have their youngsters ready to emerge from the nest shelter in time to take advantage of the pollen-filled staminate cones on the pines. These clustered cones, guarded only by paper-thin scales, offer fats and proteins that provide important dietary supplements to the food constituents in the soft cambium tissue of young twigs. The young squirrels creep out to the branch tips to enjoy the fresh products of the trees, and often tumble clumsily only to catch themselves or to drop unharmed all the way to the ground. The heavy-bodied adults must be more cautious as they cling with their back legs and reach out with forepaws and jaws.

Until the end of summer, the tassel-ears show no indication of the long hairs for which they are known. Then the young of the year grow thick tufts, while the older individuals still have no tufts or only scanty ones by September. Slowly, however, during the winter while the animals' fur coats protect them from the cold, the ear hairs grow longer and this age difference disappears. By mating season the tufts are tassels twice as long as the ear itself. Extending for an inch and a half or somewhat more, they give the squirrel an air of aristocracy. But soon after the young are born and by mid-June at the latest, the tassels are shed. Their loss is the final step in the spring molt, which begins when the winter snow melts and daily grooming gradually clears away the heavy fur insulation over the body.

By trading altitude for latitude, the tassel-ears encounter reasonably uniform weather conditions over their complete range. Yet the squirrel populations have been isolated by geologic changes over the past millenia into three unequal parts. Between the Durango squirrels in Mexico and their kin in the United States, a desert without ponderosa pines forms an extended barrier. Near the boundary between Arizona and Utah, an upward bulging of the land has been sliced through by the racing waters of the Colorado River, creating the Grand Canyon. This separated a small population of tassel-ears on the Kaibab Plateau, north of the Grand Canyon, from their relatives elsewhere in the western United States. None of the Kaibab tassel-ears descend into the main canyon, which is a mile deep and fourteen miles from rim to rim, or cross the tributary canyons to the east and west. Nor do they venture beyond the forest into the deserts of Utah. Isolated for millennia, the Kaibab tassel-ears *(Sciurus aberti kaibabensis)* have evolved as distinctive animals; most have dark belly fur and an all-white tail rather than the white belly fur and gray striped tail inherited by the more widespread population, called Abert squirrels *(S. aberti)* to honor the western explorer who discovered America's tassel-ears.

The Kaibab squirrels are all protected within Grand Canyon National Park and the adjacent national forest lands. Yet their numbers vary considerably, above and below an average total of about 1,000 individuals. No one yet knows why, unless each increase follows a second breeding season and second litter in years when oaks bear particularly heavy crops of acorns. As though to take advantage of such an opportunity, the males retain their sexual ardor from mid-March until late July. After weaning the first litter of three to four young, each well-nourished female seems able to repeat her performance and double the chance that the Kaibab Plateau will have all the tassel-eared squirrels it can sustain.

The Ice-Lovers

Each April in recent years, the news media have reported bloody massacres on the ice in the Gulf of St. Lawrence, where baby seals are clubbed and skinned to support the fur trade. So many claims and counterclaims center on the gore, the economic gain, and the possibility that an occasional seal might eat a fish of a commercial species, that the behavior of the seals themselves fades into the background. How much does a mother seal suffer, and for how long, if she sees her baby killed and separated rudely from its soft, white pelt? Some day, and perhaps sooner than we anticipate, the heavy hand of human exploitation may be lifted from these animals that come by age-old habit to haul out on the ice and bear their young.

Scientists have their own jargon for these seals: they are "pagophilic pinnipeds," meaning ice-loving fin-feet. The hind feet of seals and walruses are specialized as swimming organs, held together and sculled vertically for propulsion in the water. The feet drag along on ice or land, where locomotion for these mammals is so much more difficult and slow.

Five different kinds of seals in the Far North follow the ice-loving habit. Most abundant are the harp seals *(Pagophilus groenlandicus)* of the North Atlantic and Arctic Oceans, the ringed seal *(Pusa hispida)*, and the solitary bearded seal *(Erignathus barbatus)* of the Arctic. Rarest is the ribbon seal *(Histriophoca fasciata)* of the far northern Pacific Ocean. One population of the common harbor seal *(Phoca vitulina)* in the general vicinity of Bering Strait follows a similar pattern. All these animals time their annual return from the sea to the ice shelves as an innate compromise with immense survival value. They arrive to bear their pups and to breed not in relation to the greatest extension of the ice, which is from January until early March, but when spring is beginning to separate the ice from the land and to fragment the shelf into floating floes. Then the danger from native predators, such as wolves and foxes, that might come from shore is at a minimum. Unfortunately, human predators in boats can arrive whenever they find it profitable.

The schedule the seals follow allows less leeway than most in nature. The bulls (males) of various ages arrive first and establish their territories. In those seal species that are sociable breeders, the competition for frontage along the edge of the ice shelf becomes intense. Divisions of space are still being fought over when the cows come, almost synchronously, a few days later. Each pregnant female has an inherited program for this annual rendezvous, as near as possible to the site on the ice shelf where she herself was born. She returns after eleven months away, with a baby to deliver. She hauls out on the crumbling ice instead of on land, as most kinds of seals do. Her labor begins almost immediately. In the sociable species, her efforts are watched over by a harem master—the bull who will impregnate her again without delay. She will continue to nurse her baby, if it survives, for about a month on the ice. By then the sun will melt the support and force all of the seals into the cold water. The young must be independent or perish. Not until another winter renews their preferred support will the pagophiles again have an opportunity of this kind.

In parts of the Arctic where seals of more than one kind use the ice shelves simultaneously, each of the gregarious species tends to cluster by itself. All of them bear young with an extraordinarily heavy coat of fine fur that serves as important insulation. It conserves heat in the youngster's body, and also keeps the body heat from melting the underlying ice and producing a puddle. Pregnant ringed seals give their babies some extra protection by seeking separate pockets in the rough surface of the ice, and leaving their newborn pups in these partly concealed lairs. Skua gulls and jaegers generally fly down to scavenge for each afterbirth a mother seal expels. Sometimes the birds come so close that both the mother and her pup rear up in self-defense.

From his extensive observations on seals of several different kinds in the Arctic, Dr. Ian A. McLaren of Dalhousie University in Halifax, Canada, has concluded that the females of solitary species are the most efficient mothers. Unlike the social seals that accept life in a crowded harem, they have a choice of breeding sites and benefit from experience in their ice-shelf environment. As they age, they succeed in getting a larger percentage of their young to independence and seem to grow older themselves at a slower rate. By contrast, the females of social breeders mature quickly and have little chance to show that they have learned anything. Quite often their pups get crushed by the harem bull as he clumsily hurries to mount a cow or to drive off a competing male who has crossed the territorial boundary.

The smaller kinds of ice-loving seals have no real opportunity to give their young extended care, for their stay together on the breeding shelf is brief and a pup can scarcely follow its mother after they enter the water because of the storms and turbulent seas so customary in high latitudes. Although the mother ordinarily is pregnant again, the embryo developing inside her causes her no inconvenience.

Seals follow the pattern of development known as delayed implantation, in which mating occurs at a convenient season but the embryo is kept at an early stage, unattached and floating in the narrow central cavity as though in warm storage. Later it will burrow into the tissue of the uterine wall where the mother's blood brings nourishment, and complete the major changes in growth that lead to birth on schedule. For an animal in which gestation requires only three or four months, this program seems the best way to have the bearing of young and the reimpregnation of the mother occur one after the other in the most propitious month of the year.

By far the largest of the ice-loving fin-feet is the walrus *(Odobenus rosmarus)*, whose name is a corruption of the Scandinavian for "whale-horse." A bull walrus may be sixteen feet long and weigh 3,000 pounds, whereas a full-grown cow rarely is more than eight feet in length or weighs above 2,000 pounds. After she attains maturity at an age of four or five years, she attracts the attention of each bull she passes. Generally she chooses one male to follow, and becomes a member of his small pod, or harem. This does not make life much easier for her, for she often gets scarred in his affectionate but clumsy embraces.

The actual mating antics of walruses have rarely been observed because the animals court and copulate in the cold water amid small floating ice floes or on larger masses of drifting ice. Mating occurs on northward migration, at almost any season between February and June, depending on geography. Walruses travel south beyond the limit of fixed ice each winter, from Arctic waters in the Pacific as far down as the Pribilof Islands of Alaska and in the Atlantic occasionally to Iceland and Newfoundland. Along the western shores of northern Europe they need not swim so far because of the warming effect of the Gulf Stream.

To produce a calf four feet long, weighing between 100 and 150 pounds at birth, the mother walrus nourishes it inside her body for eleven months or more. For a year and a half to two years after it is born, she will nurse her single youngster. At first it has no teeth and hardly enough fatty insulation to keep warm in the cold water. She encourages it to use its front flippers to hold her by the neck while she swims at the surface with her baby on her back, riding along above water. When she dives for food, the baby goes along, clinging tightly and holding its breath.

At two years of age, the young walrus has tusks from two to four inches long and knows how to use them as its parents do. It can dig into the sea bottom with them while head down, almost vertical, at a depth as great as 200 feet. Freeing bivalve mollusks, large snails, worms, and sea stars in this way, the walrus uses its flexible lips and bristly moustache to push the food into its broad mouth for chewing and swallowing. The walrus can use its tusks equally well to haul itself out headfirst on floating ice. Not until a young bull is about seven years old, however, will its tusks be large enough to use in contests to settle the social hierarchy. Then a few individuals will win the right to a small harem, while the losers go off into bachelor herds. The bachelors stay separate even during migration, when great mixed herds of dominant bulls and their small pods of cows and calves endeavor to keep together.

The Great White Bears

Where the drifting floes and scattered icebergs form an indefinite southern boundary between the shelf ice and the open ocean farther from the North Pole, the seals and young walruses often encounter the most marine of all bears—the polar bear. Formally known as *Thalarctos maritimus* ("the sea bear of the sea"), it is a streamlined swimmer that never hesitates to enter the icy water. Dog-paddling with its huge front feet or propelling itself with all four feet at once, the bear crosses open stretches twenty miles or more in width. In this way it reaches some of the icebergs it can see floating near the horizon.

Male polar bears, often eleven feet long and weighing 1,700 pounds, and those females that are not pregnant usually remain active all winter despite storms. They creep over the ice toward the wary seals, relying upon the camouflage of their white coats to get within rushing distance. Alternatively, the bear swims below the ice to a seal's breathing hole, which is too small for its broad shoulders. But by jarring the ice, the bear alarms any nearby seals. Although unable to see any danger, the seals plunge down through the breathing hole. The waiting bear has a good chance to seize a seal before it can get away. This stratagem pays off more frequently than might be expected.

Pregnant polar bears seek out denning areas in October. These are either in regions of pressure ice or along hilly parts of polar coasts. Pressure ice consists of great confused masses that have been piled up by

currents driven by gale-sized winds and then frozen solidly together. Snow, although usually scanty in the Arctic, accumulates in drifts on the lee side of ice masses and in the valleys. The female bears dig dens in the snow with one or two rooms large enough to turn around in, connected to the outside world by a narrow entranceway. The next storm may fill the entrance passage except for a small hole kept open by the melting action of warm air from the breathing of the big bear inside. She drowses there until early December, when her one or two cubs are born. Afterwards she is more alert, nursing them and keeping them clean without emerging from the den. Not until March or April will she lead them forth onto the sea ice and break her long fast by hunting again.

Like other bear cubs, the newborn polar bears seem extraordinarily small, naked, and helpless. Each weighs less than two pounds, compared to their mother's 700 pounds. They keep warm by snuggling into the long white fur between her front legs, where her nipples are located. Nourished only by her rich milk, they grow rapidly. Within a week their fuzzy coats of baby fur appear. At six weeks their eyes open. Yet for another six weeks at least the cubs will have no chance to look around at the polar seascape or landscape adjacent to the den. They will depend largely on their mother's milk until they are nearly ten months of age, when they will begin to learn from her in the open how to get other foods for themselves.

By the quirky month of April, the mother polar bear begins to rebuff her year-old cubs and encourage them to go off independently. Weighing about 200 pounds each, they can manage fairly well. The time has come for her to regain her attraction for adult males of her kind—that is, males that are five years old or more. Generally she does not stand on ceremony, but accepts the first healthy male that comes along. For a few days she responds to his cautious courtship, then sends him on his way. Like her rejected cubs, adults must hunt independently to wring a satisfactory living from the polar territory, where hardships are the rule.

Sociable Little Whales with Teeth

From time to time in summer, along sea coasts in the Far North, polar bears feast on the carcasses of small, sociable whales that have perished after stranding themselves. The largest of these creatures, known variously as pilot whales, pothead whales, or blackfish (*Globicephala* spp.), prove to be the scarce males. They may be twenty feet long and weigh about three tons. Females are much more numerous, but rarely exceed sixteen feet in length as measured from the tip of the rounded snout to the notch between their tail flukes. Both sexes are sleek, with a glossy black skin that may be marked with a white area below the chin.

Polar bear (*Thalarctos Maritimus*)

The bulging head of a pilot whale tends to distract attention from its closed mouth. The animal snaps open its jaws only to seize prey, which generally is an eighteen-inch arrow squid *(Illex)* of almost cosmopolitan distribution. Each victim is held by small, strong teeth, of which the whale has seven to eleven above and below on each side. It can also use them to eat cod and other bottom-frequenting fishes above the continental shelves if the supply of squid gives out. This event must be rare, for fishermen seldom complain that pilot whales are interfering with their livelihood.

More often the fishermen interfere with the pilot whales, waiting until a substantial number are near a coast and then driving them ashore with boats. Each whale yields on the average about 40 gallons of blubber oil, 2 gallons of head and jaw oil, and an abundance of meat that can be used to feed mink or foxes on fur farms. In Newfoundland alone, between 3,000 and 4,000 of these whales are caught annually for the shore-based whaling industry.

Groups of pilot whales are generally too large to call a pod. Instead, they form herds composed of as many as a thousand individuals, all going in approximately the same direction and within communicating distance of one another. In summer they travel where their favorite squid are most abundant, which is from coastal waters to the open ocean but chiefly between latitude 40° and latitude 60° north or south of the equator. They reach these latitudes soon after the spring equinox—a few weeks before their month-long mating season is due to begin.

The logistics of family life among pilot whales seem as strange as the frequency with which these animals strand themselves. It is the males, and not the females, who are least ready to breed. Females show almost none of the coyness ordinarily attributed to members of their sex. They attain sexual maturity by six or seven years of age, and seem ready to mate at almost any season. The ovaries in each female produce egg after egg from puberty until her age exceeds eighteen years, unless she is pregnant or nursing a calf. Males mature more slowly, not reaching full sexual development until their age is twelve years or more and their bodies are almost sixteen feet long. Since the mortality rate among males is twice that among females, relatively few males live to breeding age. Those that survive appear unable to mate except during a month or two each year. Most of their sexual activity is concentrated in April and May in the Northern Hemisphere and from October to November in the Southern Hemisphere.

Now that a few pilot whales have been maintained in huge marine aquaria, the natural sounds these mammals make in their underwater world are being recorded, studied, and played back in attempts to learn more of their communication system. A knowledge of sounds from the captive pilot whales will help scientists recognize those from a free herd among the confusion of calls in the open sea. The calls back and forth while a sexually aroused male is making his rounds among the many females provide special interest. Almost certainly he vocalizes while using his bulging forehead (which whalers call a "melon") to butt against a potential mate. No actual fighting has been noticed in these herds. But males that have lived through several breeding seasons ordinarily bear many scars on their fins. Such scars may record underwater battles.

After the mating season ends, the mature males leave the main herds of females and young pilot whales. Instead they associate with a few others of their sex and age, plus a small number of full-grown females. Almost always these are aging cows, past their reproductive years but far too alert to be called senile. A study of the dentine layers in the teeth of these big animals in small herds reveals that some of both sexes may be 50 years old. The male continues to participate in reproduction annually to the end of his life. But for a female to stay healthy and active for more than 30 years after she passes through the climacteric is astonishing indeed. Probably this prolongation of the natural lifespan parallels the pattern of longevity in humans. It reminds us of the interpretation made by Dr. G. P. Bidder, one of the first British scientists to study human aging. He suggested that "no man ever reached 60 years of age until language attained such importance in the equipment of the species that long experience became valuable in men who could neither fight nor hunt." Throughout the animal kingdom, in almost every instance where old, nonreproductive individuals live on, their presence in some way helps protect members of the breeding population. Among pilot whales, we can almost be sure that the old females protect the mature males between seasons of polygamous activity by sharing with them the hazards all adults face.

A pregnant pilot whale carries the calf inside her for between 15½ and 16 months of fetal development. When the youngster is born, usually in mid-August in the Northern Hemisphere, the mother is already in warm waters near the tropics. The calf will be eight or nine feet long, larger if a male. It will have about six months to depend entirely on its mother's milk before its teeth push through the gums and she starts her seasonal trip toward higher latitudes. The growing youngster follows her for about 22 months before it is weaned and can become just another juvenile in the herd. Each sequence of pregnancy and suckling occupies almost 40 months of the mother's life, and limits her matings to one in four years —no more than three in a lifetime. It is easy to understand why she takes no chance in rebuffing a male by being coy!

The herds of pilot whales in North Pacific waters and around the world at the latitudes of Tasmania and New Zealand are often accompanied by a smaller sociable whale—the bottlenosed dolphin *(Tursiops truncatus)*. This is the amiable creature that our friend Dr. John C. Lilly has popularized in his book *Man and Dolphin* and other writings. This particular species possesses a wonderful repertoire of underwater clicks and calls, with which it communicates among members of its kind and echolocates for food. It cooperates with animal trainers and experimenters, seems extraordinarily playful and willing to imitate, and reacts resourcefully to events below water and above. At maturity it has a brain that is bigger than ours, although with many similarities. Whether it is more intelligent seems open to debate—after all, it is man who is trying to teach dolphins to serve him, not the other way around.

The jawline of a bottlenosed dolphin bears an anthropomorphic resemblance to a human smile. The facial expression remains when the animal opens its mouth and shows the 20 to 26 teeth on each side, above and below. For fishes and squids and even some of the larger shrimp this is no joke; each dolphin eats a quarter of its own weight of these prey animals every day. Yet for the dolphins, catching food is a form of social game. Generally five or more dolphins cooperate in driving their victims into a compact group or toward shore where escape is difficult. One dolphin at a time darts forward to claim a mouthful, and returns to swallow and herd while each other dolphin eats in turn.

Bottlenosed dolphins probably range more widely than has yet been discovered. During winter in the Northern Hemisphere they are not uncommon in the Mediterranean Sea and the Bay of Biscay, as well as along the Atlantic coast of North America. For the summer they spread northward to the Baltic Sea and British waters, and to the Bay of Fundy between Maine and the tip of Nova Scotia. In the Gulf of Mexico they seem satisfied to stay without migrating. But a separate population in the Southern Hemisphere follows a schedule of traveling according to seasons similar to the travels of those in the North Atlantic, although six months out of phase. The dolphins visit New Zealand and Tasmanian coastlines chiefly between January and June.

While traveling toward higher latitudes each year, the migrating males put on a special display of aquatic acrobatics, swimming upside down below the females they are courting and emitting a wonderful assortment of calls that may start as solo yelps but turn into duets as the pair respond to one another. Some males reach this state of excitement in March, although a good many don't until May; for most, April is the customary month in the North Atlantic. A closely related dolphin in the North Pacific follows the same schedule.

A pregnant dolphin carries her single baby for almost a year before giving birth. Toward the end of her pregnancy she enlists the cooperation of another female, who stays close by and helps during the emergence of the baby. Regularly the tail comes out first, which seems to keep the newborn dolphin from attempting to breathe until it is free of its mother and can be nudged to the surface of the water. All whales are probably born in the same amazing way.

Both the mother and her assistant make sure that the newborn dolphin gets to the surface immediately. They keep it there until it can learn to take a deep breath, close the blowhole atop its head, and submerge for a short distance. At first, the baby cannot hold its breath for more than 30 seconds. Then the time lengthens. In just a few hours the mother is able to begin suckling her youngster each time it leaves the water surface. The next day her routine has settled into a more efficient one: suckling for about fifteen minutes continuously, and letting the baby follow her around (breathing as it needs to) for half an hour before giving it more nourishment. By then the baby has adopted her way of sleeping: floating about a foot below the water surface and awakening at fifteen-minute intervals, just in time to swim upward and refill its lungs with air.

A newborn bottlenose has its eyes open immediately. Its body is about three feet long and weighs about 25 pounds, as compared to the eight- or nine-foot length of its mother and her 400 to 600 pounds. The young dolphin almost doubles its dimensions during the first month after birth. Often it tries to eat bits of the fish and squid that it sees its mother taking so hungrily. But it simply cannot stomach these substitutes for milk. It gets nauseous and vomits up everything it has eaten. In another week or so, it tries again. Not until it is six weeks old will it have any teeth. It will be on schedule if it weans itself before the end of its fourth month. Some youngsters refuse to do so until they are a year and a half old.

Gradually the intimate association between the mother and her youngster fades into the general pattern of sociable reactions with which dolphins adjust to one another. The seeking of physical contact and the playful imitations typical of the nursling mature into a readiness for teamwork, whether in catching food or in exuberant swimming. Games that resemble tag and follow-the-leader provide daily practice in precision swimming, and also in leaping into the air, usually to return to the water with scarcely a splash. All these maneuvers are accompanied by outbursts of submarine sound with which the animals coordinate their activities.

Occasionally a bottlenose dolphin makes an attempt to include people among its playmates. One example a few years ago began among the fishing boats in the harbor of Opononi, New Zealand. First the dolphin let fishermen scratch her gently with an oar. Then she swam so close to children in shallow water that they could rub and stroke her. The dolphin made a special friend of a thirteen-year-old girl, Jill Baker, and soon allowed the girl to ride for short distances as though on an underwater pony. The animal invented a game with a colorful beach ball that floated. Sometimes more than a thousand people gathered on the beach to watch the porpoise play with the children and the ball. The summer idyll ended in March, 1956, when the seemingly intelligent animal allowed herself to become stranded by a receding tide. With no one present to give assistance, the porpoise died of suffocation, unable to breathe when not buoyed up by water.

The self-stranding of both pilot whales and bottlenose dolphins remains inexplicable but far from rare. No evidence has yet been found for any poisoning that might upset the normal alertness of the animals, nor for the close presence of predators such as the killer whale, *Orcinus orca,* a cosmopolitan relative with particularly voracious habits. The phenomenon has a seasonal quality as viewed from shore. In northern waters it is a quirk of April more than of any other month, but is still a mystery.

CHAPTER THREE

The Rhythms of May

MAY HOLDS SPECIAL significance for mammals of many kinds, ranging from the meadow jumping mouse of North America to the African impala. Whatever their geographic range, they must begin reproduction at this time of year if their young are to benefit later from the most suitable resources the habitat offers. Sometimes near kin with similar environmental requirements follow unlike schedules as a result of differences in the time required to get the young to weaning age and the gradual accumulation of adaptive features whereby competition between species is minimized and diversity sustained.

The Meadow Jumping Mouse

Late one afternoon in mid-May we sat motionless by the mill pond in our New England town, watching for animal activity along the boundary between the dry world and the wet. A few toads were still trilling in the shallows. Beyond, where water-lily leaves had recently reached the surface, a big bullfrog tried out his bass voice. He was the first of his kind we had seen or heard that season, and we wondered how long it would take him to get his full mating call into operation.

Suddenly a small mouse with a tremendously long tail dropped out of the air within inches of the pond and crouched on the wet sand as though stunned. "Meadow jumping mouse," one of us whispered. "*Zapus hud-sonius,*" the other responded, "the big-footed mouse of the Hudsonian life zone." For more than thirty seconds the three-inch rodent with the four-inch tail made no move that we could see, except to twitch its short black whiskers. The elongated toes on its hind foot reached almost to the wrist of its front foot. Then its small ears cocked forward a little, and the mouse crept to the water edge to lower its pointed nose and drink. Soon satisfied, it turned around in a jerky, nervous way, walked along a few inches until its tail was straight out behind, and leaped past us like a stone from a slingshot. Landing somewhere in the grass fully four feet from its takeoff point, the jumping mouse was gone. Its abrupt mode of locomotion leaves no trails, and often must save its life. A frightened rodent of this kind has been known to jump ten feet on level ground, and jump again a fraction of a second later. For its size, it is one of the most expert jumpers in the world. Only the barn owl seems able to outmaneuver a meadow jumping mouse and catch it.

Usually this small rodent moves about at night in some damp grassland, between Alaska and Labrador, as far south as the Carolinas. Its home in winter is a carefully constructed nest. During the active summer season, the mouse lives alone or with a companion in a dome-shaped mound of plant fibers woven tightly together. Generally this one-room house is well hidden within a clump of grass or below a bush. Those we find seem to have an entranceway along or below a loose rock or fallen log. If we lift the rock or log, the mice rush out and leap away with froglike jumps too fast to follow.

For most of every night all through their active season, the jumping mice scurry around in search of food. They pay almost no attention to rain, and swim across streams or small ponds that form part of their home territories. Individual mice vary greatly in the extent of their travels. Some rarely go more than forty yards from their summer nest in any direction. Others seem to patrol areas as great as two and a half acres. These variations permit each acre of suitable territory to support an average of three mice in the spring and ten in the fall.

From late spring to fall, the diet of the meadow jumping mice changes gradually. Until midsummer, animal foods predominate. Caterpillars and beetles account for about half the total intake, and seeds only about a fifth. As summer ends, seeds and fruits increase in importance. Subterranean fungi largely replace insects. No doubt this shift from high-protein to high-carbohydrate diet reflects the availability of acceptable foods. But it also matches well the needs of the jumping mice. The proteins from animal foods quickly rescue mice that have fasted for many months in hibernation, and the carbohydrates supply the kind of nourishment that is most easily converted to fat for the next period of dormancy.

At our latitude—about 43° north—the meadow jumping mice hibernate from mid-October until late April. Toward the end of September they begin hunting for a well-drained hillside or a sloping cliff face into which they can tunnel for as much as six feet. Each mouse prepares its own nest at the inner end of its tunnel, where the earth itself will provide good insulation against cold. Only an occasional jumping mouse drags some fungi or dry fruit into the sleeping chamber. Most of these animals seem to enter hibernation with no snack near by and with an empty digestive tract. Instead, they rely upon body fat to supply their modest requirements until spring. Less than a quarter of an ounce per mouse will suffice.

Astonishingly, only about a third of the meadow jumping mice that enter hibernation survive to eat again. Two-thirds die of starvation, seemingly because they did not prepare themselves adequately for six months or more without food. This seems to explain why the average weight of the mice that emerge in the spring is actually a little more than the average weight at the beginning of hibernation.

So far, no one has discovered which mice get fat enough and which do not. Our own suspicion is that those jumping mice born late in the summer never really have a chance. We know that the matings in May lead to a peak of births in June. At three weeks of age, the five to six youngsters

Meadow jumping mouse (*Zapus hudsonius*)

in a litter get their eyes and ears wide open. At six weeks, they are fully grown and on their own. They have until autumn to get fat. But their mother often accepts another suitor, and bears a second litter in August or September. Except at the southern edge of the range, where summer lasts longer, these youngsters of the second litter may go into hibernation before they are physically ready. And far south, a third litter sometimes arrives before the reproductive and feeding year comes to an end. What opportunity do these latecomers have to grow up and fatten before cold weather comes?

Our interpretation of evolution and genetics requires us to assume that each structural and behavioral feature of any living thing has a definite survival value now or did at some time in the past, or is an unimportant byproduct of some process that does (or did) have statistical significance for survival. We feel uneasy about dismissing as inconsequential a mortality rate of more than 60 percent from a single cause, even if jumping mice are in no danger of becoming extinct because of this annual loss.

If our suspicion that the jumping mice born in late summer are the ones that starve in winter is correct, we would focus on the readiness of the mice to mate a second and third time during each season of activity as the inherited basis of the high mortality. For zero population growth or any reasonable excess to furnish potential colonists and the raw material for future evolution, one litter annually might well suffice. Two litters might suffice where the warm weather releases the mice from hibernation earlier and lets them find food longer in the fall—and where barn owls are more common hazards than they are close to the Canadian border. Perhaps this almost fruitless effort toward increasing the population of jumping mice is a pre-adaptation; evolutionists sometimes list other features in this category that prove wasteful of lives and effort until some change in weather or a geographic barrier gives the needed opportunity to those creatures that are predisposed to benefit.

One other feature of these long-tailed mice, and of their woodland counterparts in the Northeast, challenges our understanding. Not only do the males have a baculum or penis bone almost a half inch long (about the same size as their extended intromittent organ), but the females have a homologous clitoris bone. In a female meadow jumping mouse, the clitoris bone is about a fourteenth of an inch long. The woodland jumping mouse *(Napaeozapus insignis),* which is the smaller of the two species in our area, has a clitoris bone a tenth of an inch long. What benefit these bones confer in the genitals of the little mice is beyond our imagination.

Bacula have long puzzled anatomists. Why should a raccoon, or a dog of the same body size, have one four inches long, while a male bobcat does not? Does a walrus have a special need for a penis bone the size of a baseball bat? Why do a few men, particularly of ethnic groups from the Middle East, have a penis bone? They may be totally unaware of this possession, although it shows up in an X-ray of the pelvic region or in a postmortem examination. This curious bone formation persists as a feature found in members of several different mammalian orders.

Long-term Hibernating Ground Squirrels

To the west of the Appalachian foothills and into high country and north country almost to the Pacific coast live burrow-making members of the squirrel family whose schedule of hibernation seems extreme by most standards. Either drought or cold sends them underground by mid-August or early September, and keeps them there for nine or ten months every year. The approximate date of emergence from dormancy seems governed by the timing of spring thaw. When the snow melts and the ground softens, these animals pop out. After a few meals have taken the edge off their hunger, they court and copulate. Not a day can be wasted if their young are to complete their embryonic development, be born, grow to independence, attain full size, and fatten for the foodless months ahead.

By contrast with our northeastern groundhog, which can devote the first month in each season of activity to finding a mate and consummating the relationship, the yellow-bellied marmot *(Marmota flaviventris)* of the Middle West and lower slopes of the Rocky Mountains can spare less than two weeks. The hoary marmot *(M. caligata),* the loudest whistler on the high slopes and above timberline, must perform even more quickly. In ten days or less the males efficiently make their rounds, leaving every female of their kind pregnant except those less than one year old. For their part, the females have no time to be coy and choosy. In the next months, their intake of food must be enough for their own needs plus the litter of young developing inside their bodies and nourished after birth on their milk. Litters vary from three to eight, but average between four and five. To satisfy her own needs and theirs, the mother will browse from dawn until half an hour after sunset every day, so long as she sees no danger and hears no neighbor whistle sharply in alarm.

Smaller relatives of the marmots—the Columbian ground squirrel of rocky slopes and adjacent lowlands in the Northwest, and the striped ground squirrel of the Great Plains and contiguous grasslands—have a similarly short season for activity between long months of hibernation. The Columbian ground squirrel *(Citellus columbianus)* is a sociable vegetarian, whose safety from hungry badgers and bears improves where immovably heavy rocks guard the entrance to its burrow. It gets into trouble when it attempts to colonize the grasslands nearby at lower altitudes, where winter is shorter but man is an intolerant competitor. The striped ground squirrel (or thirteen-lined squirrel, *C. tridecemlineatus*), which American explorers named out of patriotic fervor when the flag had thirteen stripes and thirteen stars, is more of a predator. Its lengthwise stripes and rows of squarish spots blend with the grasses while it stalks and pounces on grasshoppers, caterpillars, mice, and prairie bird nests. Although it rounds out this diet with seeds, fruits, and the tender leaves of grasses, it destroys so many harmful insects, rodents, and weeds that it improves man's pastures and meadowlands. Only on agricultural crop-land does it become a pest.

We have often sat quietly among colonies of Columbian ground squir-rels in the high valleys of Montana, where the rules of Glacier National Park protect them from the stockmen. The nearest squirrels generally stood stiffly upright, about a foot tall, beside their entranceways, making no sound that we could hear. They were sentinels, and unsure of our intentions. Beyond the danger zone we could see other squirrels of this kind nibbling away on glacier lilies, wild onions, and other herbs and grasses. More than 90 percent of their diet consists of vegetation. The Columbian ground squirrel telescopes a whole year's feeding activity into two or three months. The only concessions to life under these conditions seem to be made by the young of the year: they risk the autumn snow-storms to feed a few weeks longer than the adults, and spend less time digging a deep burrow with many chambers beyond the reach of frost, so long as they can find food in the open.

When we think of their timetable, we cannot help but be impressed. Not until mid-May does the snow ordinarily melt and evaporate enough to expose the entrances of the ground squirrels' burrows. The animals must wait for this just inside the plug of earth with which they sealed the doorway late in the previous summer. Out come the adults, some weeks ahead of the still-immature young from last year. After a stretch or two, the full-grown animals are off to find whatever plants have started to grow. Old males need the least food, for they alone have been provident enough the previous summer to store a collection of seeds and bulbs before going to sleep. While waiting for springtime to release them from their hibernating dens, these old males have dined in bed. After a few good meals outdoors, perhaps to restore their vitamins, these males will cast caution to the winds and go courting. They pursue every female they find.

Each female Columbian ground squirrel can accept a male for a few minutes, almost like a participatory floor show or a dance between courses in her extended meal. That she is promptly pregnant need not immedi-ately concern her. Not until three weeks later will she need to prepare a special nursery den, larger than her hibernating room, with a soft mat-tress of fine dry grasses. There, about 24 days after mating, she will give birth to from two to seven naked, blind, toothless babies, each weighing between a quarter and a third of an ounce. On her milk they will increase their weight fivefold within a week, open their eyes and begin to chirp at three weeks of age, be weaned at four weeks, and independent at five. They have less than two months more in which to grow up, start digging a winter home, and get fat enough to survive.

Drought in late summer, causing the fresh vegetation to wither, is less of a hazard to Columbian ground squirrels that live on the mountain slopes than to those on the lowland plains or to the striped ground squirrels of the grasslands. Grasses are well adapted to drought, ripening their grains and dying if they are annuals or going dormant—but surviving in the roots—if they are perennials. The insects match this schedule by growing while the grassland plants are green, and then waiting in a dormancy known as diapause until the green growth starts again. The dormant stage of the insect may be an egg, an immature individual, a pupa, or an adult. It will be out of sight, and hard for a striped squirrel to find. Birds that have raised their young will be ready to fly south for the winter season. The striped squirrels may still find mice, but must compete for them with hungry predators that would accept a squirrel for dinner just as readily. If the squirrel is fat enough when prolonged drought causes this shrinkage in the food supply, it simply retires below ground, plugs the doorway, and lets its body temperature sag to the level characteristic of hibernation. The month may be August, barely half-way through summer. For the squirrel, it might as well be winter.

The next year for the striped squirrels begins between mid-March and the end of April, when the adult males awaken from their deep sleep. The females ordinarily do not appear until two or three weeks later. Until April or May, hunger seems predominant and the stripers are all busy hunting. Then the females begin repairing their burrows, cleaning them out or digging new nests. For a month or more they seem ready to mate. Strangely, it is the males that show disinterest; each becomes sexually active on his own schedule and for only a week or two.

By the end of May, courtship ends and the pregnant females prepare to produce and raise their annual litter. It may include ten babies, each weighing less than a tenth of an ounce at birth. Within 12 days their nakedness has been hidden under a fur coat with the full quota of stripes and spots. At 28 days, their eyes open and their bodies weigh three-quarters of an ounce. At about 35 days they creep out of the burrow for the first time, with another week in which to grow before they are weaned and must leave home to seek their fortunes. Those squirrels that survive for 90 days are fully grown, and already hibernating in their own chambers. In half as many months as a chipmunk the same size needs for activity each year, the striped ground squirrels fit their predatory way of life to the prairie climate. They survive the winter in the groundhog's way, without going to all the trouble of filling subterranean chambers with stored food in the compulsive pattern of the chipmunks.

Jackrabbits: Theme and Variations

Alert as any ground squirrel, the long-eared jackrabbits of the American West sit on their haunches with hind legs tensed and ears almost vertical, testing the crisp dry air for every sound. Through binoculars we can see the nose tip quiver, the whiskers and eyelashes move, and the ears revolve slowly like radar antennae scanning the sky for incoming aircraft. These are animals of open country and exposed mountain slopes, where they can leap along at 45 miles per hour to escape a predator. Yet on their own, they rarely travel more than a quarter mile from where they were born. Home, for them, is usually less than 40 acres.

The most widely distributed of the jackrabbits are the black-tails, which are distinguished from the more northerly white-tails by having a black stripe from the tip of the four-inch tail along the top of the tail and up the adjacent region of the back. Ranging from Washington state to the isthmus of Tehuantepec in Mexico, and east to the 100th parallel of longitude, they include the giants of all jackrabbits—the antelope jacks *(Lepus alleni* and *L. gaillardi)*, which often attain a weight of thirteen pounds in the southwestern deserts—and the medium-sized California jackrabbit *(L. californicus)*, which rarely grows to weigh more than half as much. It has gray sides rather than white ones, although this is difficult to see on an overcast day or when the sun is low, the principal times when the California jackrabbit feeds in daylight.

From the northern to the southern extremities of their range, the California jacks encounter great variations in climate and progressively longer snow-free seasons southward. Through the natural processes of selection, they seem to have assessed the local opportunities for raising young. It is as though they allowed for the length of the growing season, as measured from the average date of the last killing frost in spring to the average date of the first killing frost in autumn, divided the number of weeks by six, and decided on how many litters to have. In the northern part of their range, where the growing season may be less than twelve weeks long, the California jackrabbits have one big litter. Farther south they produce two or three in quick succession. At the southern end of their geographic distribution, they fit in four or more. Yet, within each region, the jackrabbits also seem able to adjust their reproductive activities in relation to the size of their population. As the number of jackrabbits in an area increases, the time of the first matings in the year comes later (as it does northward), while the number of young in a litter diminishes (as it does southward).

Sheep raisers are interested less in whether California jackrabbits choose May or April for starting the season's productivity than in how many of these native animals are present and how many sheep could be accommodated if they were exterminated. For every sheep on Arizona's rangelands, there must be 30 jackrabbits fewer. On the salt deserts of Utah, the comparison changes according to the season. In spring, 10.2 California jackrabbits could live on what each sheep needs for food. In winter, only 5.8 need be present to consume the amount of forage suitable for a sheep in an equivalent time. Even the choice of foods is extraordinarily similar. In early spring both sheep and jackrabbits nibble the buds from shadscale and whitesage, and supplement this diet with tender herbaceous plants and young grass blades as these appear. By late spring and through the summer, grasses hold the greatest appeal. But as the grasses mature and dry, the jackrabbits turn again to browsing on the shrubs, adding quantities of saltsage and big sagebrush to their menu.

As we think of the California jackrabbits persistently trying to raise their young in so many areas where only sheep of all man's domesticated animals can eke out a living, we wonder about the complexities of competition. What else could use the same vegetation? To what foods will their predators change if the population of jackrabbits is reduced drastically? For millions of years the coyote and bobcat have relied upon catching jackrabbits. We feel sure that an undefended lamb or a starving sheep, or any animal (wild or domestic) that has died and become carrion, will be only a partial substitute. Yet if the predator is eliminated along with the jackrabbits, the alternative prey (such as mice, ground squirrels, and grasshoppers) can build up much larger populations than before. Each change affects the environment that man covets for his flocks of sheep.

The jackrabbits use their range for more months of the year than many of their neighbors, wild or tame. Unlike the marmots and ground squirrels, they take off no time for hibernation, and give the vegetation no respite from their feeding. The jackrabbit is likely to be there when man brings his sheep, and still there when he takes them away to better pastures or to the slaughterhouse. Yet this is the schedule to which the plants adjusted long ago.

To learn about jackrabbits in their natural state, we have often sought out small airstrips where the grass is mowed occasionally. There we can sit, partly concealed, until the jackrabbits that live on the limited grassland forget our presence. Ordinarily a mother jack spends the sunny hours in one of her saucer-shaped depressions (known as a form), slightly camouflaged by her own drab color and by the plants that grow on all sides. When she has young to bear, she gives birth in a form. Although their eyes are open and they could follow her, she usually keeps them quiet where they were born for a day or two, additionally concealed by fur that she pulls from her belly skin. Then she leads them out one at a time, and hides each youngster in a different place. As we watch, we see her hop from one to the next, nurse it, and move on. Between their meals of milk, the young are already nibbling on the nearest vegetation. Soon they are weaned and independent, and the mother accepts another male.

Most of the male jackrabbits remain ready to mate at almost any time of year. Their show of courtship seems casual: a little sniffing, a few nose rubs, sometimes a moment of urination on the prospective partner. Copulation usually waits until after sunset, and by morning the male has ended his liaison. The females seemingly accept these brief attentions in any month, but few do so after September or before January. Most fail to become pregnant before spring, wherever they live. When May comes, however, we can count on finding every healthy female jackrabbit either pregnant or with newborn kits to tend as the consequences of a courtship. Throughout the range of the black-tailed jackrabbits, no other month is so fruitful, nor are litters so large as in May.

Bouncy Antelopes Near the African Equator

If our field experiences in many lands have taught us anything, it is not to expect natural events in the temperate zones to prove similar to those in the tropics. We cannot assume that all months are alike to tropical animals and plants just because the sun is at the zenith twice each year and never casts much shadow obliquely at noon whatever the month. Length of day and night change little with the seasons, but the seasons themselves are real because of the varying direction of the sun. The great convection patterns in the atmosphere, produced by solar heat, move north until the summer solstice and then south until late December. The plants react to changes in moisture, and the animals to variations in the amount of greenery.

Many of the conspicuous mammals of Africa migrate on an annual schedule where the high plains extend north of the equator to the fringes of the great deserts, and in the opposite direction to the nearest parts of South Africa and South West Africa. Generally the animals spread out during the dry season over areas with abundant surface water and tall native perennial grasses. When the rainy season begins, the animals travel to regions with temporary rainwater pools and annual grasses to bear their young. Zebras and wildebeests follow this program. So, to a slightly lesser extent, do the springbok and gemsbok, eland, and giraffe.

The travelling wildebeests are the ones familiar to crossword-puzzle fans as gnus, or more specifically as brindled gnus. The zoologist knows these animals as *Connochaetes taurinus,* and may recall that in 1850 they were assigned to a separate genus *Gorgon,* which is now regarded as unnecessary. We find amusement in this reference to classical Greek mythology, implying that the wildebeest is as terribly ugly as a Gorgon. The live animal does seem to lack the grace we recognize in almost every other kind of antelope. Its face is narrow and oxlike, with bristly hair and beard.

Its legs are long and slender, and its tail so long-haired at the tip that it drags except when the wildebeest is bounding along with a hobby-horse gait. At intervals the animal stops and turns its head to gaze behind, as though to see if it still is being followed. It is easy to think of the wildebeest as some sort of joke.

Actually, the wildebeest is a successful experiment in evolution. Some 13,000 of them live in Kruger National Park in South Africa, where they outnumber all other conspicuous animals except the 28-pound impalas and the 475-pound zebras. The total weight of wildebeests per square mile is exceeded only by the impalas, zebras, 1100-pound buffaloes, 2500-pound hippopotamuses, and 7000-pound elephants. By almost any measure, wildebeests are a success. Similar numbers of them roam in the national parks of equatorial Africa. These herds merely represent the uncounted millions that traveled in this part of the world a century ago.

Until the past decade, no one understood the behavior of wildebeests. Now the situation is clearer because observers using modern techniques have shown that at least three different social groups comprise the herds. The mature cows and their calves are the real travelers. During regular migrations they pass through the ill-defined territories of dominant bulls, which remain in restricted locations for as much as two years at the peak of their prowess. On the most desirable terrain, the bulls are less than 100 feet apart. On less suitable but more average areas, the distance between one bull and the next is closer to 200 feet. Beyond these, on the least advantageous sites, are bachelor herds that move about almost continuously; some of these animals are young and healthy, but they have not yet succeeded in displacing a dominant bull from his holding. Actually, real battles rarely develop. Instead, the bulls go through an elaborate show of beginning belligerency and bluff their competitors into moving elsewhere.

Probably the bulls are willing to mate at any time of year. They have a chance to do so only between April and June, principally in May, toward the end of the wet season. This opportunity, which identifies a rutting period, comes as the cows and their calves migrate toward the area they will live in during the dry season. Each cow comes into heat seemingly as a consequence of all the attention she receives from one bull after another. The bulls try to detain the mature cows and their female calves, and commonly each bull becomes the center of about 25 members of the opposite sex. However, each bull does his best to chase away any male calves, despite the efforts of the mothers to protect their young of either sex. Eventually the male calves do leave their mothers and seek the quieter environment of a herd of bachelors.

The cows and female calves continue their migration, only to come into the territory of the next bull. There they are courted again and delayed, serving for a few days as members of another pseudo-harem. By the time

the grazing grounds for dry season are reached, the females have spent a while with perhaps a dozen bulls, all of which have copulated with them several times. About 37 percent of the calves that are 16 months old and around 83 percent of the young cows 28 months old become pregnant; virtually all of the older cows become pregnant regularly. It is impossible to tell which bull has made any of them pregnant.

The timing of these progressive orgies fits well into the annual cycle of changes in vegetation on the African savannas. The bull wildebeests have had an opportunity to graze for most of the wet season on the best greenery their limited territories can produce; each animal should be at the peak of health and virility when the migrating females begin to arrive. And when the pregnancies that ensue come to full term eight to eight and a half months later, the females will be back in their wet-season area just in time to benefit from the fresh growth. It will nourish them during their months of nursing and give them the strength to migrate again, meeting the expectant bulls on another lap of the course.

The chief obstacle to untangling these unlike patterns of behavior in the bulls and cows has been that both sexes look so much alike. No one expected them to act so differently. The heavy, upturned horns and goatees are worn by cows as well as bulls. Even the animals themselves seem to have difficulty telling the sexes apart. Apparently they do so by scent rather than by sight. A cow that detects a threatening gesture from another wildebeest nearby discharges a little urine, whereas a bull does not. From the odor of the urine, the other animal seems to learn her sex, perhaps her age, and certainly her reproductive condition.

The principal predators on wildebeests—lions and hyenas—probably also discover by scent which sex they are near. The lions tend to concentrate on the more stationary bulls, whereas the hyenas follow the cows and calves. The calves are the most vulnerable. On the average, four out of five of them fail to survive their first year. Although they can run along beside their mothers within a few hours after being born, they must make no mistakes because a mother will nurse only her own calf. If the calf gets lost among the herd, the mother seems to make no effort to find it. And if the calf does not reach its mother in a day or two, it weakens from starvation and lags behind until the predators kill it. The larger the herd is, the more likely the calves are to get lost. Dense populations also favor the spread of cattle plague, called rinderpest, a virus disease to which young wildebeests are particularly susceptible. In both ways the size of the population tends to stabilize itself.

The behavior of the migratory cows is often the most conspicuous feature of wildebeests. Where food is abundant, the animals and the calves attending them tend to disperse. They regroup when threatened by predators. Commonly the wildebeests associate temporarily with zebras, seemingly to mutual advantage. The zebras have better vision, and keep scan-

Wildebeest (*Connochaetes taurinus*)

ning the horizon to detect danger while there is still time to move away. The wildebeests are more likely to recognize the presence of predators by scent, and to warn the zebras by nervous actions.

When food is scarce, the wildebeests move on in their own distinctive way. Long lines, perhaps a dozen animals wide, travel tirelessly across the land. They spend each night in the same formation, and return to it as quickly as possible if they are panicked into dispersing by an unexpected attack. Only during the rutting season, when progress is delayed repeatedly by the territorial bulls, does a wildebeest population gain a temporary stability.

The Graceful, Vulnerable Impala

Some of the wildebeests' neighbors become occupied with mating at the same time of year, but without migrating and with far less conspicuous social organization. May is the season for sexual activities among the graceful antelopes known as impalas *(Aepyceros melampus)* commonly found among open thornscrub and the acacia savannas from Uganda and Kenya to South Africa. The long, lyre-shaped horns on the bucks let us recognize them instantly; the ewes lack horns and are a trifle smaller. Both are equally sleek, glossy, reddish brown above and pale tan below, and distinctively marked with tufts of black hair a few inches above the hoofs on the hind legs.

Impalas make prodigious leaps over shrubs, rocks, each other, and sometimes just a narrow paved road. When running, their successive jumps may span 20 to 30 feet. But they seldom run far, and avoid open country. As much as possible they disappear among the undergrowth in a way that reminds us of deer in the Northern Hemisphere. Like the deer, impala are favored by the regeneration of forest land that has been cleared and abandoned, and by the invasion of brush into grasslands that have been overgrazed by domesticated animals. They thrive in country too poor for cattle, so long as they have cover, food, and a source for a daily drink of water. Unlike deer, however, they share their range with many other kinds of hoofed animals. They automatically minimize competition with these different neighbors by preferring food plants that the others usually pass by. An impala takes mostly grasses, and will make the widespread Bermuda grass 50 percent of its diet if it is available.

Impalas settle most of their family disputes in April. The bucks dare one another with shaking heads and loud, hoarse grunts. Most of this display gives males of different sizes and degrees of vigor a chance to judge which is the more powerful, and to move apart without a fight. Less often the competitors are almost evenly matched and a battle develops, continuing until one individual yields. By the end of April the winners have managed to fragment the loose mixed herds into smaller groups of 15 to 25 instead of hundreds. Each group then consists either of ewes and lambs led by a dominant buck, or of a similar number of bachelors with lower status.

The lambs, which have quite recently been weaned, provide no interest for the rutting buck. His attention is sufficiently divided between checking each ewe in his herd several times a day, to learn by scent and taste whether she is about to become receptive, and making sure that no lesser buck lures away any female from his harem. June may arrive before every ewe is pregnant and can be allowed to wander off. He will have lost interest in all females as such long before the six-month pregnancies conclude, almost always with the birth of a single lamb. By then the loose structure of the herd will have re-established itself.

At the forest edge and under isolated trees, impalas provide fairly easy targets for leopards crouching upon overhanging limbs. Undergrowth aids prides of lions in ambushing these antelopes. Wild dogs in packs pursue individuals relentlessly until they fall from exhaustion and can be consumed. Jackals and hyenas often kill impala lambs. Even a chimpanzee or a baboon will eat a lamb if it is slow in leaping away. Together the predators harvest a wealth of nourishment from the impala herds without destroying them. Somehow, births restore the population. Yet these kills, which seem absurdly easy to a person watching from a parked automobile in an African national park, prove to be more selective than might be guessed. The young males are particularly vulnerable, perhaps more so than the lambs and oldsters. We have never been able to see why this is true, although we can appreciate its significance. Each buck less means that much more grass for the survivors and that many fewer contests for dominance in April. When May arrives, each hundred impalas actually needs only four to six bucks to father the next generation.

CHAPTER FOUR

When June Brings Summer to Alaska

RARELY DO FRIENDS returning from touring Alaska tell us of encounters with giant brown bears, the largest carnivores on earth, with giant moose, or with the northern fur seals, which the Alaskan coast has more of than any other part of the Arctic. None of these are evident from the rail of a ship sailing the Inside Passage, or from the Alcan Highway, or from an airplane flying to a polar airport.

To meet these spectacular mammals we must go where few people venture. June offers the best advantages for doing so, since this month brings the Far North its longest days. The winter snow is melting, nourishing green plants and helping them grow rapidly in the continuous light. Surprisingly little of the melt water runs away. Rainwater is similarly retained; the permafrost under the thin soil resists erosion that might deepen any stream channel. Pools and ponds form, reflecting the sky overhead. Other places seem just as wet underfoot because they are covered more than ankle-deep with soaked peat moss (*Sphagnum*) of pastel hues ranging from yellow-green to brown and red.

The Brown Bears of Alaska

Alaska has many thousands of square miles of back country in which the brown bears roam. Largest of all are the Kodiak bears (*Ursus middendorffi*) of Kodiak Island and adjacent smaller islands. The grizzlies *(U. arctos)* are slightly smaller. These are North American members of a circumpolar species known in Europe simply as the "brown bear." Strangely, the Eurasian bears can be tamed easily and are used in circuses, where they dance and show off amiably. North American grizzlies are individualists, far too unpredictable to be domesticated to this degree.

By June in Alaska, the solitary male bears are seeking the company of adult females, or sows, unaccompanied by young. At this season they wander widely, and cease to follow the straight pathways they use so regularly after the middle of July. When amorously inclined, brown bears stop observing their careful habit of placing their front feet where some other bear has stepped, and then their hind feet in the spots from which they have just lifted their front ones. Tracks of this kind less than ten years old may just be linear patterns of holes ten inches in diameter, five to ten inches deep. Older tracks become furrows, often going up and down hill without swerving to either side, as though they had been laid out by a surveying team.

The feet of the average male make a print two or three times as broad as those of the average female. His weight will be about 1,000 pounds, and hers around 700. In June, a female with no family will tolerate his approach. For a while he may be satisfied to graze beside her. His shoulders and head come close to hers as the two bears stuff coarse reed grass and horsetails into their mouths, chew, and swallow. He may let her step ahead while he sniffs at her body. He can test her sexual readiness by resting one big paw atop her rump. If she does not whirl and beat him off, snarling to emphasize her rejection, he is likely to mount her and copulate for several minutes. Then they resume feeding side by side. Sometimes the huge animals rear up, facing one another, and wrestle ponderously.

No one can predict what will happen if a second male joins the placid pair. One or the other may drive him off. Or the first male may amble away, seemingly unwilling to battle a bear as big or bigger than himself. Or the mated pair may accept the stranger and form a threesome, with each male standing politely aside while the other enjoys the female's favors! In this respect, the big brown bears take their time. An average copulatory episode lasts 23 minutes; some continue for a whole hour. In her season, a female may stand for six or seven episodes before she rejects all further advances.

For the male grizzlies, the mating season begins in May and lasts until early July because individual females come into heat during different portions of this time. For each female, the period of readiness is rarely as long as four weeks. During this period, however, her hormones prepare her to be affable to passing males twice, with a lapse of from four to eighteen days between mating sessions. The male that consummates courtship with her during her first session is likely to be elsewhere when she becomes receptive the second time. For this reason, a female grizzly usually mates with two suitors in a season. Up to four males per season is not unusual. One young female grizzly that the Craighead brothers and Maurice Hornocker followed year after year had a total of eleven suitors that she accepted in various years without showing any preferences.

Staying close to an Alaskan brown bear for days or weeks can be a risky undertaking, but consecutive observation yields far more toward understanding the animals than facts on separate bears pieced together from a random sample. The persistent scientists who study grizzlies or Kodiak bears are less likely to kill one than licensed hunters, just as the hunters are less likely to conduct a detailed and knowledgeable postmortem examination of the trophies they take. For this reason, no one is certain (although the probability is high) that a fertilized egg goes through only preliminary development before reaching a dormant stage. Only part of the 210- to 245-day interval between the last copulation and birth should be necessary to produce a newborn cub the size of a rat, or two to four cubs (the average number is more than two), for that matter.

Births take place inconspicuously while the mother bear is in her winter den. She may not rouse herself and emerge until March or April, when the cubs are two or three months old. For a while, the remnants of winter will be on all sides and the adult bear must use her power to claw open the ground and expose edible roots. If she can find carrion at this season,

Grizzly bear (*Ursus arctos*)

or discover a colony of hibernating ground squirrels, she feasts while her unweaned cubs play about her or rest quietly. The young bears grow slowly, but she guards them well. They will share her den a second winter and sometimes a third before she weans them and sends them off on their own. Then she is free to breed again.

The advantages of such parental care are evident in some census information about Kodiak bears in the Karluk Lake drainage area. Licensed trophy hunters are permitted to take a limited number of full-grown bears, but select the biggest males as far as possible. Their success appears to explain why female adults outnumber males about fourteen to five, even though the two sexes are equally numerous among the younger age groups. Cubs less than a year old constitute 26 percent of the population; yearlings, 22 percent. Those that are sent off on their own do not find food so regularly, and their numbers diminish: two-year-olds account for only 17 percent of the population, three-year-olds 10 percent, and mature bears (some not yet fully grown) 4 years old and older a mere 25 percent.

To keep so huge an animal satiated requires more food than most giant bears can find. They are almost continuously hungry, challenged by their environment to fill their large stomachs and relatively short intestines. No longer does the Far North furnish these animals with enough caribou, moose, musk oxen, or elk. Nor has man been willing to share his cattle, sheep, hogs, and horses with the bears. Consequently, over immense areas, America's biggest carnivores have either vanished or are turning increasingly to a vegetarian diet. On their present range they have to subsist on plant food for most of the year due to lack of prey or carrion. A caribou or moose that is injured or diseased may go down under the bear's high-speed rush, but these opportunities come unpredictably and seldom. When they do, the carnivore gorges itself and caches the remainder. Then, rather than trust so wonderful a windfall to the camouflage of earth, rocks, branches, and leaf litter that the bear has pulled over the carcass, the animal lies down watchfully close by and prevents any scavenger from depriving it of its next meal.

The almost chronic hunger of the big bears can be judged from the regularity with which they disperse themselves over the countryside when elderberries ripen in August. Using their big paws like hands, they shove whole clusters of fruit into their mouths, chewing and swallowing every bit. Day after day the bears feast until the elderberry supply gives out. Then they head back for the rivers and lake shores, where salmon may be coming to spawn.

Never is the agility of the brown bears so impressive as when they are darting into the riffles and lake margins, pouncing on the alert, slithery fish. Many a man has tried to be as efficient in coordinating eyes and hands and been frustrated, where a grizzly or a Kodiak bear would scoop out a three-foot salmon, carry it proudly to shore, and shake it to death. Often the excitement of catching fish appears to become a sport, for the bears continue with one catch after another without taking time to eat. When the big predator does settle down to a meal, it may be unable to hold all the fish it has caught. Observers from the fisheries industry are particularly resentful of the bears when any big salmon "go to waste." Actually, nothing is wasted in nature, for the smaller scavengers and decomposers have relied upon this largesse for countless years.

For many of the grizzlies, the feasting on salmon puts an end to the active year. The nights have begun to lengthen noticeably. The migratory birds have gone. Frosts come more frequently and any precipitation is likely to be snow. The time has come to rehabilitate the hibernation den or to excavate a new one. Then a week or two of wandering may convince the bear that it might as well retire. Into its den it goes, curling up as much as its size and bulk allow. Breakfast can wait until spring.

Whether dozing in its den or lumbering across the tundra, a grizzly stimulates our imagination to piece together what we have learned about the Ice Age—the Pleistocene. It was then or just previously that ancestors of modern bears entered the New World from the Old. They crossed from Siberia to Alaska by way of a low-lying corridor of land perhaps 200 miles broad, where the salty waters of the Bering and Chukchi seas now roil back and forth over a shallow bottom. In open country the big brown bears found an acceptable mixture of animal and plant foods all the way to Mexico, and across the Great Plains to the forest edge along the western slopes of the Appalachians. Their descendents stayed in the open until competition with invading people became too great. The last survivors in Mexico vanished less than a decade ago, half a century after California lost the last wild grizzly to represent the animal on the state flag. Now fewer than 850 can be found in the conterminous United States, all of them in the national parks of Montana and Wyoming except for a few in shrinking wildernesses of Idaho and the San Juan range of southwestern Colorado. Since 1968 the grizzly has been on the list of rare and endangered species compiled by the United States Department of the Interior.

The Black Bears of American Forests

Forest-inhabiting bears from the Old World found a large realm for themselves from coast to coast in America, where they became the familiar black bear *(Euarctos americanus)*, with outliers in Florida and northern Mexico but not in Newfoundland or the West Indies. Unlike other bears, they climb trees readily and regularly send their cubs into the topmost branches if danger threatens on the ground. Like the grizzlies, black bears hibernate in a deep sleep generally unaffected by sound or light. Yet they can awaken and defend themselves vigorously without noticeable delay if pricked or pinched in any way. This reaction must have been essential for survival where smaller predators, such as cougars and wolves, were once fairly common and especially hungry in winter. A bear could rout any of them from the narrow doorway to its den, and could easily hold off a pack.

No one knows how many environmental factors enter into the computerlike nervous pathways of a mature bear to be matched against the inherited program in its memory bank and to settle the timing of hibernation. The sex of the bear and its condition (such as pregnancy) certainly enter into this automatic computation. So, apparently, do abundance of food, sunlight, temperature, and precipitation—particularly snowfall. In the northern woods and at higher elevations on mountainsides, these factors and probably many others combine to send black bears into hibernation in October and keep them there until early May, when the males get restless. A pregnant female starts sooner and is likely to stay longer. In the south, each bear remains only a few days or a week in hibernation before emerging to stretch, find something to eat, and go to bed again.

The inner clock by which the bear's life is timed strikes for the pregnant female sometime in January. She awakens wherever she is, at least enough to help her cubs as they are born. If this is her first family, a single cub can be expected. In later years she is likely to have twins—one for each nipple. Occasionally triplets appear. Quadruplets are rare. One mother bear has been seen with five cubs at her heels, but whether they were quintuplets or included some that she had adopted—unable to distinguish them from her own—could not be determined. By the time she leads them from the den they weigh about four pounds each. While she dozed, awakening only at intervals to freshen the bedding, the cubs have grown to eight or nine times their weight at birth.

After the mother and her family leave the den, they do not return to it before autumn. Within their territory they can find many places to sleep, either in a shallow depression scooped by the mother from the soft earth at the middle of some thicket, or in a tree, where she can relax completely stretched out on a horizontal limb four inches or more in diameter, her legs hanging down limply on the two sides. Her biggest problem is the restlessness of her cubs, who want to play whenever they are awake. Within limits, she tolerates their foolishness even if it interferes with her sleep. But when they exhaust her patience, she cuffs them with one paw, at which they cower down whimpering and remain quiet for a few minutes.

Black bear cubs grow bigger and mature faster than do brown bear cubs. By autumn of their first year or by the end of that year's hibernation at the latest, their mother rejects them and gets ready to mate again. Generally her young of the previous season stay together for their second summer. Then they too become solitary, seemingly in preparation for an awareness of the opposite sex, which develops by the time they are three years old.

As we compare the black bear of America with the circumpolar bear-of-bears, *Ursus arctos,* the similarities impress us more than the obvious differences. The fundamental pattern that grizzlies showed when they too were part of the living community from Canada to Mexico revealed the same flexibility to suit climatic differences. The blacks take advantage of the leeway that is inherent in their program of embryonic development, as it is in all bears, by harboring their fertilized eggs in a state of delayed implantation for a month less than the grizzlies and two months less than the polar bear. This they accomplish by waiting until mid-June or even July to mate, as though there were no hurry.

Not even man has been able to hurry the black bears noticeably. In their independent way they have accommodated themselves to his presence. In his national parks they have learned to beg for handouts, while their grizzly cousins have merely discovered how to rake garbage from the park dump. So far, no intelligence test devised by scientists can reliably rank the various bears according to their antics, or allow a convincing comparison between bears and other kinds of wildlife. But woodsmen and photographers who associate with these animals by choice and without firearms assure us that any bear merits respect.

The Northern Fur Seals

A scientist's first reaction to an unfamiliar animal often gets perpetuated in the name given it. Thus the northern fur seal became *Callorhinus ursinus,* literally "the little bear with the beautiful nose." Its velvety pelt finds wider appreciation. In the cold waters of the North Pacific Ocean, as on the islands of the far North where the seals haul out between May and August, these animals were persecuted to the verge of extinction. Only in 1911, after the original herds totaling perhaps five million had been reduced to fewer than 130,000 fur seals, could any international agreement be reached on a Fur Seal Treaty to protect the survivors. It called for an immediate end to commercial killing of seals at sea, and a harvest on the breeding islands limited to males shown to be superfluous in an annual census of the population. The recovery of the herds to between one and two million animals has been gratifying, and praised as "the most outstanding of all accomplishments in the conservation of wildlife." The managers of this living resource have learned a lot about seals in the process, but a good many mysteries remain.

The females swim to shore in June from as much as 3,000 miles away in the open ocean, often to bear their young in the same rookery area where they themselves were born a few years earlier. The pregnant individuals arrive a few weeks after the males, which rarely stray as far as the females between one breeding season and the next. Each cow seal carefully chooses a landing place along the shore. She goes where she can see other seals of her kind, and yet avoids the hauling grounds adjoining the rookery on each side, where only nonbreeding seals congregate. In some way she decides against the full-grown bulls, ten years old or more, who remain mateless until August even though they hold territory either just above the high-tide mark or on higher ground. Instead, she goes to some younger bull with a territory some 75 to 250 feet in each direction and joins his harem—usually one that is already started. She is accepted eagerly, as though a prodigal wife who has just returned.

The mother-to-be may have from 2 to 48 hours in which to rest ashore before giving birth to her single pup. Attentively she nurses it and listens to its squeaky voice. To her ears that sound must become so distinctive that she can later find her pup by listening, even though it is sprawled among dozens or hundreds of others of the same size and appearance. The cow then submits to the bull seal, who has been waiting impatiently for this moment. He is black, with a gray cape, and weighs as much as 300 pounds. She is gray, and not over 140 pounds. In their turbulent, boisterous courtship and copulation they may easily brush aside or stand on (and crush) her newborn pup, which is glossy black and weighs a mere 12 pounds. If the pup survives, it receives another meal before the mother goes off into the sea to relieve her hunger. One side of her forked womb recovers from harboring the pup while the other side becomes the home of her next youngster.

Usually the mother seal returns in a day or less to exactly the same spot along the shore. She scrambles out as best she can and searches anxiously for her pup. She will nurse no other, although all the hungry squeaking youngsters try to hold her attention. Around and through their squirming bodies she finds her way, listening for the one call that she recognizes. From June until August she repeats this search every time she comes ashore. Not until it is two months old will the youngster be ready to follow her into the water and learn to catch its own food.

For the dominant bulls, the departure of the rookery population must be a great relief. These males have not had a bite to eat or anything to relieve their thirst for three active months. During all this time they have stayed on their posts and battled to keep females in their harem from being abducted across the boundary lines. Their fat has shrunk and left their pelts flabby. Fresh sores, scabs, and old scars show that some of their battles have been violent. Now these bulls have earned a nine-month respite in the sea, eating to restore their size and vigor, resting or just rolling in the waves.

At the rookery, however, new activity has begun. Out of the ocean have

come virgin females—seals slightly more than two years old. The full-grown bulls who have had no mates all summer welcome the fresh arrivals. So do some of the younger seals who have entered the rookery from its inland periphery to claim territories abandoned by the dominant bulls. In the melee of competition for the young female seals, the rookery assumes a wild confusion quite different from the crowded conditions earlier in the summer. Now there are no pups lying about, waiting to be nursed. Soon the mating season ends, and the shore seems deserted. Hardly a seal will be in sight when the first snowflakes begin to fall.

From the rookeries on St. Paul and St. George Islands and Sea Lion Rock in the Pribilof Archipelago to the west of the Alaskan mainland, between a third of a million and more than a half million northern fur seal pups swim away each August to seek their fortunes in a life that may last twenty years. Of all those born, about nine out of every ten launch themselves, while scavengers clean up the remains of the one pup in ten that dies on the rookery rocks. Of dead pups in good enough condition to reveal the cause of death, more than a fourth have died of starvation—generally because their mothers did not find them or failed entirely to return to the helpless pup. Slightly more died of hookworm disease.

In addition to the natural death of pups on the Pribilof rookeries, which represents a loss to the population of between 25,000 and 75,000 each year, the authorized representatives of the fur industry have been taking an average of nearly 50,000 adult males in the age groups from two to five—chiefly three-year-olds. At these ages the body is between 30 and 51 inches long, and the pelt has not yet been scarred by territorial battling. To avoid damaging the pelts, these males are driven inland from the hauling grounds and relieved of their skins as quickly as possible. Any two-year-olds that are too small or five-year-olds that are too large for their pelt to have a market value are allowed to scramble back to the sea.

In June 1969 the census team on the Pribilof Islands counted 2,342 territorial males with females on the rookery areas, and 7,935 that as yet had no harems. Three weeks later, in mid-July, the total number of territorial males had increased slightly, and 7,385 had females while 3,212 did not. The number of females can be inferred from the 303,500 pups born that summer—one pup to each pregnant female. Herding 303,500 mother seals into 7,385 harems is a monumental task for the big bull seals, but it averages out at 41 cows per bull. Some harems contained as few as five cows. No one seems to have counted how many the real sultans sequester to themselves! The truth resembles a tale out of the Arabian Nights, under the low moon of the Alaskan sky at the end of June.

Nourishment for so many seals—those born each year, plus those born the year before that have not yet returned to Alaskan waters, plus the older seals that return to the hauling grounds and rookeries—requires a supply of fishes and squids weighing at least ten times as much as all the

Northern fur seal (*Callorhinus ursinus*)

seals combined. Commercial fishermen are interested in knowing what kinds of saleable fish the northern fur seals eat. Scientists of the National Marine Fisheries Service, in the United States Department of Commerce, conclude that the seals rely upon whatever they can find at night in surface waters, and that the northern anchovy *(Engraulis mordax)* constitutes more than a third of the diet. Rockfishes of the genus *Sebastodes* and capelin *(Mallotus villosus)* together account for fully a half. Saleable fishes, chiefly salmon, are among the remaining one-sixth, along with several kinds of squid.

Sea Lions of the North Pacific

A male Steller sea lion *(Eumetopias jubata)* may weigh 2,500 pounds, and each of his mates nearly 800 pounds. These animals appear from cold waters of the North Pacific Ocean in late May and June to establish rookeries of their own on coasts and islands from Japan to the Bering Sea and southward to southern California. Even though the world has no more than 150,000 Steller sea lions and their pelts are of little value, the food requirements of so many big animals could cause concern in the fisheries industry if saleable fishes constituted much of the diet.

Again, the scientists of the National Marine Fisheries Service have provided the answers. The Steller sea lion chooses capelin in Alaskan waters, plus rockfishes, sand lances, sculpins, and flatfishes if they are unable to escape. Off the California coast, these sea lions seem to concentrate on flatfishes and rock fishes, for which they dive as much as 300 feet to hunt along the bottom. They range as far as 85 miles from the coast, but more often congregate within 15 miles of land. Despite the fact that the stomach of a Steller sea lion may contain a mass of fish weighing nearly a tenth as much as its whole body, the big sea-going mammal rarely seems to take a commercial fish or to interfere with a net in any way. They do, however, patrol the nets and catch lampreys that are attacking the fishermen's haul.

The one opportunity we have had to meet some families of Steller sea lions on a rookery was just off the California coast at Santa Barbara. It came just a few days after we had approached in a small skiff at Monterey a barking mob of the smaller California sea lions *(Zalophus californianus)* that were basking on a jetty. From our angle, scarcely above sea level, and at a distance so close that we could smell the fish on their breath, the sleek-skinned muscular bodies of the California sea lions seemed dangerously big. The largest male weighed no more than 200 pounds. How much more huge the Steller sea lions seemed when we clambered across the stones toward them! Instead of barking, the 2,000-pound bulls bellowed a deep-throated, prolonged roar at us. They rolled their eyes as though not quite willing to turn their huge bodies, they seemed ready to drive us from their beach. Then we discovered that they were far less courageous than the females who had pups to protect. If we approached too close, the bulls hurried into the water; their belligerence was only active toward other males of the same kind, who might displace them as harem masters.

The California sea lions are far harder to approach on their breeding areas, for they choose shorelines where precipitous cliffs rise above the water and caves have been undercut by storms. Within these sanctuaries they bear their young and mate, seemingly endangered whenever the waves break vigorously at high tide. Yet these are the animals that have adjusted so easily to captivity that they are shown off as trained seals, doing tricks in return for a herring or two as a reward. We suspect that they would outdo themselves if their trainers gave them squid or octopus instead, for whenever California sea lions have a choice, they prefer the cephalopod mollusks as prey rather than any commercial fish. Curiously, all sea lions that are caught ashore are apt to have empty stomachs. Apparently it is their custom to digest what meals they catch before emerging from the water.

Surely it is more than a coincidence that the northern fur seals, the Steller sea lions, and the California sea lions choose June for the pregnant females to haul out, bear their pups, and mate with the dominant bulls. They follow this routine all the way up and down the coast of the North Pacific Ocean. It must be an inherited habit that transcends the variations among species. It is not connected with the schedule their young will follow in moving to the water and weaning themselves on seafoods instead of milk; this varies from one kind of sea lion to another, as does the final size and bulk to which the animal will grow. Even the northern fur seals that colonized some islands off the California coast a few years ago, as though proving that their northern breeding grounds were no longer adequate, followed the same tradition by bearing their pups in June.

CHAPTER FIVE

The Slack Month of July

IN THE WORLD of nature, the progress of most animals and plants either gains its annual momentum during June or is slowing for the summer as June arrives. July throughout the world tends to be a slack month, when few enterprises are begun. Close to the equator it may be wet season or dry, depending on the location. North of the tropics it is definitely summer, and south of the tropics a time of increasing chill.

For those mammals that do wait until July or even August to seek a mate, more than this behavior makes them seem odd. Yet when the lives of each kind are examined methodically, the appropriateness of their timing becomes evident.

The Nine-Banded Armadillo

Among all the many mammals in America, only the nine-banded armadillo *(Dasypus novemcinctus)* seems to wait until July for mating. It is apparently unique in regularly producing four identical quadruplets from each fertilized egg. Their development does not begin, however, until many months after the parents meet. Actual gestation takes only about 120 days, and may end with birth anytime from February to April. Just a few hours after they are born, the youngsters trot along after their mother like miniature facsimiles.

Armadillo (*Dasypus novemicinctus*)

The first armadillo of this kind we met was crossing a Texan field soon after dawn. The animal was big, weighing almost fifteen pounds and measuring sixteen inches from the tip of its pointed snout to the base of its sturdy tail. To avoid being scratched by the stout claws on the animal's feet, we grasped its tail as firmly as we could and kept the hind legs from touching the ground. An armadillo is almost helpless if its weight hangs mostly from its tail. Otherwise the animal can take advantage of the overlapping narrow bands between the horny shield covering its shoulders and the one over its rump to curl its body and resist.

The horny armor has a reptilian character and tends to conceal the warmth of the animal underneath. The soft skin of the underside, however, is clearly mammalian, with sparse yellowish hair, four nipples from the mammary glands, and a temperature higher than the usual surroundings.

So far we have not discovered how to tell a male armadillo from a female. Probably the animals themselves learn by scent. Their eyesight is poor, and their hearing not much better, although their thin, unarmored outer ears project conspicuously.

Our second nine-banded armadillo proved to be less than half grown. It was hunting for food through the undergrowth in Panama, grunting like a miniature pig. After glimpsing enough of its armor to be sure of the animal's identity, we pounced with both hands. To our surprise, we found its horny covering to be fairly soft. It does not harden until the animal attains almost adult dimensions, for an armadillo has no way to shed its plates and produce new ones as it grows. The youngster struggled vigorously to get away, then quieted down. By the time we freed it half an hour later, it seemed to have accepted being carried, as though it would be a good pet. It made no attempt to run off, but busied itself at once digging into an ant nest. Whenever a dozen ants rushed from a ruined gallery, the slender tongue of the armadillo slithered out, captured the insects on the sticky surface, and returned to the mouth with the load of food.

An armadillo seems to have no difficulty using its active tongue to capture and eat termites, cockroaches, immature and adult beetles, slow-moving grasshoppers and crickets, millipedes and centipedes, and spiders that pause at the wrong moment. Occasionally it takes a small toad, frog, or salamander. Any bigger prey is scarcely suitable, for the armadillo has no front teeth in either jaw, and only peglike molars, farther back, with neither roots nor enamel covering.

Such a diet is available every week in the year between the Tropic of Capricorn and the Tropic of Cancer. Nine-banded armadillos range for the full distance, from northern Argentina through Central America and Mexico. Until about a century ago, their northernmost outliers were in the vicinity of Brownsville, Texas, in the lower valley of the Rio Grande. Then human colonists settled in the Southwest and attempted to raise crops familiar to them in Europe. Armadillos found loose soil suitable for burrowing in, and attractive insect food of many kinds. By the middle 1920s, the animals spread through Texas into Louisiana and northeastern Oklahoma. Even before bridges spanned the Mississippi they managed to reach the eastern side. Perhaps with help from man, they colonized much of Florida. It was there that we met our third armadillo in the wild, after noticing it in our automobile headlight beam as it ran along a paved road.

Now we even might find these intriguing animals in Kansas or Missouri. They have expanded their range northward to the climatic barrier formed by prolonged cold weather, which keeps inactive the prey each armadillo must find to survive. It can go hungry for a few days, and escape severe cold by hiding underground. Yet it cannot hibernate, and soon starves if no insect food appears.

Despite its armor, an armadillo prefers to have the extra anonymity conferred by night before scratching around for its meals. Often several animals hunt together, and their grunts of communication can be heard for considerable distances. Soon after dawn, the armadillos either find burrows to hide in for the sunny hours or they dig new ones. Somehow several males squeeze themselves into one burrow, several females into another. Mixed sexes underground are uncommon, as though the habit of living close to others of the same sex were carried over from the womb; there four brothers or four sisters grow side by side, each set identical in every detail.

To dig, an armadillo claws at the earth with its forefeet alternately close beside its nose, and pushes the excavated material under its belly until quite a pile accumulates. Then, supporting itself on forepaws and tail, it brings both hind feet up and over the pile. With a quick kick of both hind feet at once it throws out the whole accumulation, and is ready to continue digging.

A single animal may prepare a dozen different dens, some shallow and some deep. Behaviorists suspect that the shallow, vertical excavations are pitfalls to which the armadillo can return repeatedly and find insects or other small prey waiting to be eaten. The more slanting tunnels often go four or five feet below the surface of the ground, and branch repeatedly to various chambers, each about eighteen inches in diameter. Generally several of the rooms contain a mass of bedding, consisting of grasses, weeds, and bits of shrubbery that the animal has gathered above ground. To haul the dry material down the entrance tunnel, the armadillo collects a bundle with its forefeet and pushes it into the angle between its belly and hind legs. The animal can clamp the material in place by contracting the muscles that pull on the lower edges of the armor over its rump region. Clutching its load, the armadillo creeps below ground. It rarely takes time to organize its bedding. When ready to go to sleep, it simply worms its way into the bundle that fills the chamber and curls up.

Removing an armadillo from its burrow is almost an impossibility. By stiffly extending all four legs, the animal presses its armored back against the tunnel roof. Four claws on each front foot and five on each hind foot dig into the tunnel floor. Wedged firmly in place, the creature can resist being pulled by the tail or pushed against the head.

Outside its burrow, the armadillo relies upon its speed and ability to crash unharmed through thickets, at least as a means of gaining enough time to dig itself underground. Water is no obstacle. The animal has two different ways to cross a stream or pond. If given time, it will gulp air into its stomach and become a living, floating raft below which its legs dog-paddle furiously to provide propulsion. If pressed, the armadillo simply walks down into the water, relying upon a natural lack of buoyancy to give it good traction on the bottom while it runs along as best it can. On the opposite shore the animal emerges, takes a deep breath, and hurries off. Even while engaged in exercise, it seems able to survive with almost empty lungs for underwater excursions lasting as long as fifteen minutes.

The armor for which the Spanish explorers named the armadillo ("little armored one") weights it down for underwater walking and gives flexible protection in many situations. But it seems also to get in the way at mating season. The heavy tail with its firm rings of covering cannot easily be turned aside. The short legs with their scaly coating do not spread apart readily to let a male mount a female. Instead, the male armadillo turns his mate upon her back and copulates with her belly to belly. As in all members of the order Edentata, her reproductive and urinary openings are combined. His testes are safe within his body cavity, not in an exposed scrotum. But at the critical moment he produces a penis with which to transfer his semen. Incredible as it may seem, a single mating ordinarily supplies just one sperm cell destined to find its way to an egg cell that will eventually produce quadruplets of one sex or the other.

The Spiny Insectivores

The easiest way to meet a hedgehog is to use the strategy shown to us by a friend in a suburb of London. He set out a saucer of bread and milk as soon as twilight faded, and a hedgehog came right to his doorstep to enjoy the food. The method also works well in New Zealand, where the European hedgehog has been introduced. Normally these nocturnal animals search through the gardens and beyond the hedges so inconspicuously that their presence goes unnoticed. Their quarry is chiefly insects, snails, slugs, and earthworms. They readily accept a frog, nestling bird, or mouse, or feed on carrion if they encounter these meaty items in their foraging.

If two hedgehogs arrive at the same time, attracted by the bread and milk, one of them proves its dominance by rushing at the other, squeaking and grunting, sometimes with mouth open ready to bite but more often merely to butt the submissive one until it goes away. In July and August, a similar contest transforms rapidly into a gentle nosing of one by the other, if one is an adult male and the other a female more than one year old. If she is willing, she relaxes the muscles that control the stiff barbless spines that project from the skin of her back and flanks. The male mounts her cautiously. Her spineless one-inch tail is no obstacle, and the armament on her back yields freely under his paws as he supports himself.

When any other animal disturbs a hedgehog, the muscles in its skin elevate the half-inch spines while the hedgehog folds its nose and feet against its belly and draws its spiny skin around itself. So firmly attached are the spines that the creature can be picked up by grasping a few between thumb and forefinger. A smart fox that meets a hedgehog is more likely to cover it with earth, then seize the animal on its unarmed underside when it unrolls and exposes itself.

The spines on a newborn hedgehog are soft until they dry. The youngster is born after a pregnancy of about five weeks in a litter that may number as many as seven but commonly totals fewer than five. For two or three weeks the mother keeps her young in a nest. Then their eyes open and her family can follow her about in her foraging until they wean themselves. From the outset the youngsters can climb and swim. They pull their spiny covering around them if threatened with any danger. By day they have no need to go home again, for they can hide under fallen vegetation or in holes of any kind. Yet those that live near human habitations never seem to wander far. Individuals that have been given identifying anklets return repeatedly to the same place if food is set out for them.

The Little Monogamous Deer of Eurasia

Open woodlands with thickets of undergrowth are the home of Europe's smallest deer—the roe deer *(Capreolus capreolus)*. The animal is common from Britain and Scandinavia to Korea and Japan, but is so timid and nocturnal that it is seldom seen. More often its calls are heard and mistaken for something else: a dog, when the deer barks in fright, or a nocturnal bird, when a doe screeches to summon her fawns or to tempt a buck in mating season.

Apparently we have never been in the right place on an evening in late July when, toward dusk, a roe buck begins to follow his chosen doe. She leads him through the undergrowth in a circuit, screeching at intervals as though concerned that he might be giving up. The chase grows faster. Around and around they go, wearing a track a foot wide and perhaps 50 yards long in the vegetation. Our attention has been drawn to several of these tracks, for they remain traceable for several months. Even after gamekeepers learned that roe deer made them, the secret did not get out. People were encouraged to shun the "witch circles" and to learn nothing of the edible deer that still were in the neighborhood. We too got distracted by the story and never investigated the possibility that in the center of the witch circle was the female's fawn from the previous mating. The doe returns to it after she mates, and keeps it close until it is time for her to have her next family.

Now we wonder what shape the witch circle is when the doe has twins, as happens occasionally. Is it a figure-of-eight, with two centers? The doe suckles twins separately, never together. Generally they stay where she leaves them, perhaps ten to twenty yards apart. Each fawn has about three longitudinal rows of white spots, and is a darker reddish brown than the parents.

Until late October, when the fawns are losing their spots and growing thick brownish gray winter coats like those of the adults, the roe bucks meet their single mates quite often. A buck may even engage in a second rutting period toward the end of November if by then he has not already shed his antlers for the year. More often, he has become surly and solitary if old, or sociable with other bucks if younger. When the snow gets deep, all but the old bucks form small herds of eight to twenty, under the quiet leadership of an old doe. The does with fawns also form winter herds under female guidance.

In February each buck develops the bulging buds of his new antlers upon his brow. In a month they rise to full size, although still covered by hairy skin (the "velvet"). In early April the skin dies, and drops off as "rags," and the buck polishes his armament by rubbing it vigorously against bushes. If this is his first set, his antlers will merely be single spikes close together. In his second spring they will fork, with two tines apiece of almost equal length. A three-year-old has three tines, still upright and nubbly around the base. Four and five tines are rare, and probably evidence that the buck is both superbly nourished and well supplied with sex hormones. As each buck ages, toward the average life span of sixteen years or so, his annual pair of antlers diminishes, and he may have only two tines or a single one on each side again.

As soon as a buck's antlers are clean in April, he wanders back to the area where he met his mate the year before. Since other bucks will be hunting for territory too, he has no time to waste in reaching his destination. He seems to recognize the bucks in adjacent territories and the boundaries that previously separated him from each of them. Paying scant attention unless these old neighbors trespass, he remains alert for the arrival of any new buck that is trying to acquire real estate by bluff or battle. By mid-May, when the does sort themselves out one to a territory, the roe deer have reached reasonable agreement in parceling out the woodlands. A pregnant doe can have privacy as she bears her fawn. She has contributed to her peace just a week or two before by driving away her youngster of the previous year.

Gamekeepers may seldom catch a glimpse of the roe deer on the preserves they patrol. Yet they feel concern over any dead ones they find. Of every hundred carcasses that the melting snows expose in spring, about 60 are of animals that have found too little to eat and have starved to death. Around 15 have died of diseases, some of which are more deadly for a deer that is too malnourished to resist. Another 15 have been killed by predators, with free-running dogs accounting for about half, and wild carnivores such as lynxes, wolverines, and foxes accounting for the rest. The remaining 10 out of each hundred are usually crumpled close to a road they tried to cross before an automobile struck them down. Careful examination of the individuals that have died a violent death, whether on the highway or from a carnivore's attack, show that almost 55 percent were diseased. Probably they would have died a more natural death in a few days or weeks if left to wander undisturbed—perhaps to share their infection with healthy neighbors.

By spring the surviving roe deer are ready to shed their winter coats and expose a darker color that will blend better with the shadows under the trees. They lose the distinct white throat patch and another that showed on their hindquarters during the cold months; these areas, like their legs from the ankles down, turn tawny brown instead. Gradually each deer restores the weight it lost and grows a little bigger than the year before. A buck may reach 65 pounds by fall, and a doe 35 pounds. But rarely will any roe deer reach 30 inches tall at its hips. Always it remains small and graceful, as compared to other kinds of deer.

Why should the smallest deer rut more than a month earlier than the others, and give birth at least two weeks later the following spring? Research revealed that this member of the deer family follows a schedule of delayed implantation. The fertilized egg lies dormant in the womb for about four and a half months of the ten-month gestation period. The occasional second rut in a season, in late November and early December, precedes by a week or so the renewal of embryonic growth in an egg that is already fertilized. Any offspring conceived during the second rut will be born no later than those conceived in July.

As we think about the little roe deer in their forest glades across northern Eurasia, we naturally compare them with larger deer that similarly encounter long winters and a considerable depth of snow. The same climate in the same type of country applies stress to the surviving populations of red deer *(Cervus elephus)* in Eurasia and of wapiti (or elk, *C. canadensis*) in North America. Each winter a surprisingly large proportion of the full-grown bull wapiti and red deer stags starve to death, while younger males and females of all ages survive. Apparently, after losing so much weight during the annual rutting season, which ends in mid-October, these males cannot restore their fat reserves before snowy weather is upon them. They go into the winter season relatively unprepared. By contrast, the roe buck has two extra months after his early rut to rebuild his strength. His particular schedule obviously holds important adaptive value.

Africa's Most Magnificent Antelope

When our one-time mentor on the African wildlife preserves, Charles A. W. Guggisberg of Nairobi, referred to the greater kudu *(Tragelephus capensis)* as the "most magnificent antelope," our initial reaction was surprise; offhand, we would have picked the sable antelope *(Hippotragus niger)*. But a 300-pound sable bull, no matter how beautiful his coloring or spectacular his great smoothly curving horns, cannot really compete in magnificence with a 600-pound kudu. Magnificence implies large size, not just elegance. Only the eland *(Taurotragus oryx)* is a bigger antelope than a greater kudu; and it is too obviously bovine for us to admire.

A full-grown greater kudu stands about 52 inches tall at the shoulders

as he walks along, holding his head proudly, his horn tips pointing toward the zenith. Each horn may be more than 40 inches long as measured in a straight line, or exceed 50 inches if measured around the gentle spiral. The record pair, from an animal in the Transvaal, is 51¾ inches from base to tip along a straight edge, or 71½ inches along the outside of the continuous curve.

We have met these wonderful antelopes as they move singly or in small parties in South Africa among the undergrowth of the low bushveld, and in equatorial Africa where they graze somewhat casually on the open mountain slopes of Kenya and Tanzania. Greater kudu range westward to Lake Chad and South West Africa, and northeast into Ethiopia and to the Red Sea Hills. Every wild kudu we have encountered has seemed in no hurry. This has allowed us to admire the horns on the bulls and to recognize both sexes from the four to ten wavy vertical white stripes on each flank. Otherwise the animal is mostly a reddish brown or a shade of gray that suggests the bluish tint of dry slate. A short mane continues along the back, between the prominent shoulders, and back to the sixteen-inch tail. A longer fringe of hair hangs from the throat and neck.

The greater kudus tend to herd together twice each year, for the rutting season in July and again in February and March, when the calves are born after a gestation period of about 212 days. Otherwise the full-grown bulls keep to themselves, or mingle in small groups of the same sex. A cow with her calf may stay apart from the herd, but more often she stays close to other cows, with calves and without.

At any time of year, the greater kudus excel in athletic abilities. Although none of the antelopes we have watched showed any inclination to do more than walk or lie down and chew their cud, Charlie Guggisberg and others with long experience on the African savannas assure us that kudus often leap rather than go around an obstacle. If they enter an area of thornscrub, they look over the bush that bars their path, flex their legs, and leap. They clear objects six feet high with ease and jump across rushing rivers that are fully twenty feet wide. Apparently the lesser kudu, which inhabits eastern Ethiopia and Somalia, is equally agile.

To us, the most impressive feature of the kudus is not their horns, their sedate grace despite their size, their prowess in broad jumping, or the vigor with which the biggest bulls battle in July for dominance in mating with every nearby cow in heat. Instead, it is in the subtle scheduling of their lives and their choice of food—inherited features that are as essentially kudu as the spiral twist of the horns on an adult male. This program for kudus helps them adapt to the resources their African environment contains, while minimizing competition from neighbors of unlike kinds and minimizing destruction of greenery for the benefit of all.

The timing of the rut in greater kudus, together with the seven-month gestation period, permits the season for birth to follow that of the wilde-

Greater kudu (*Tragelephus capensis*)

beests and precede that of the waterbucks. It gives the kudus a chance to benefit from whatever privacy the shrubbery affords while newborn young are getting their feet under them and learning to follow their mothers.

The choice of food is harder to pinpoint because the vegetation is far from uniform over the broad range of kudus. A complex set of factors leads a kudu to pick just certain kinds from any customary community of plants. The selection changes month by month, according to successive stages of plant growth and to the seasonal burgeoning of particular grasses and broad-leaved herbs. The array of foods that a kudu eats in any week provides marvelously balanced meals, virtually ideal in relation to its distinctive digestive system. It satisfies the animal with a minimum of actual eating and digesting, efficiently sustaining kudus at a minimum cost to the vegetation.

The preferred items for kudus are rarely those that appeal to other big herbivores in the same district. This species of antelope picks a diet that complements, rather than overlaps, the diets of all other species present. It is as though a master plan had been drawn up, allocating every part of the available vegetation to some one kind of beast—rarely two. All green plants (and some that aren't green) have a role in supporting the various herbivores. As is so often the case, inherited sensory differences and patterns of behavior effectively divide up the available resources to the benefit of a diverse fauna rather than that of any single species.

CHAPTER SIX

The Belligerent Bulls of August

LATE IN THE geological period we know as the Ice Age, primitive people in northeast Asia began to hunt progressively eastward over a chilly plain that extended where no man had ever been before. For these hunters and their families it was enough that meat animals lived there—herds of bison (*Bison* spp.) on the warmer lowlands and clusters of musk oxen *(Ovibos moschatus)* where the weather still held the chill from half-melted glaciers. These people could not have known that the plain was a land bridge to a New World, or that soon it would be inundated by the sea where the Bering Strait lies now.

Today the discoveries of scientists show that the bison had begun crossing the Bering bridge to America a few thousand years earlier, whereas the musk oxen were trying to extend their range westward into Eurasia. This difference gave the bison a slight head start in survival; they reached the Great Plains, spread out, and prospered, becoming the most numerous large animals in all of the Americas. The musk oxen were too late. Everywhere in the Old World they encountered too many human hunters and were exterminated. Before long, only some arctic islands, the Far North of Canada and the northeast rim of Greenland offered them a refuge.

The Bison's Heritage

To think of grass, bison, musk oxen and people as the complete ecological community on the Bering bridge is far too simple. Many other kinds of life were there too, including most notably the woolly elephant ("mammoth") and the wolf. Both of these animals crossed to the New World at the same time the bison did. But human hunters, while competing with the wolf for bison and other prey, eventually killed off the last of the elephants. They died where they had been feeding, along the eastern seacoast of North America. At that time the land extended from 40 to 200 miles farther east than it does today. Now fishermen foul their nets on the tusks and heavy teeth of extinct elephants along the edge of the continental shelf, and bring back trophies that reveal the past of New England almost to Cape Hatteras.

Until about two centuries ago, the wolves and bison unwittingly formed a mutual benefit society upon the Great Plains, from southern Canada to northern Mexico, from the foothills of the Rockies eastward almost to the Appalachians. The herds of bison may have totaled 40 million animals, sustained by an inherited pattern of behavior despite the wolves, the Plains Indians, and lesser predators. The wolves kept the pattern fresh by eliminating any prey animals that varied from it. The Plains Indians, afoot and with crude tools, could be tolerated because the number of bison they killed was soon restored—perhaps by supporting fewer wolves.

For the bison, life has always had a regular time for beginning but no fixed place. These great animals wander erratically, allowing the grass to recover from their grazing. But wherever they are toward the end of July, and through August into early September, the bulls and cows get restless. The biggest and most virile males, which weigh more than 3,000 pounds, try to herd the full-grown cows into harems while the cows tend to wander away. Each bull becomes suspicious of every other bull he can see. Most will move off if a big bull raises his deep voice and makes a threatening movement with his massive head. Occasionally the competitors for the right to parenthood are more evenly matched and a battle follows. From a distance of about twenty feet apart, the two bulls face each other, creating haloes of dust by pawing at the turf with their front feet and challenging each other with mighty roaring bellows that can be heard for a mile or more. Then the monsters rush toward one another, meeting head on with an impact that sounds like a thunderclap, even though it is cushioned by a thick mat of hair on the two broad brows that crash together. The sharp, curved horns are too far to the sides to have any real effect in these encounters.

Time after time the two bulls bang at one another until one has had

Bison (*Bison bison*)

enough and trots away. The victor takes a few steps as though to pursue
the loser, and then stands staring, still muttering loudly. Soon he must
round up his cows again, for they have paid no obvious attention to the
battle, and some have taken the opportunity to walk off.

The only time that a bull pays special attention to an individual cow
is when his nose tells him that within a few hours she will be ready to
receive him. He follows her closely, temporarily almost immune to distraction. If she stops, he stops—but scarcely looks away. When she moves
on, he stays right behind her. Yet he does not lose touch with the rest of
his harem. As soon as he has mated with the one cow he has been attending
so exclusively, he guides her back to the straggling group that he must
herd together once again.

After the middle of September, the bison herds move quietly, with little
commotion. On each animal the hair is getting longer, particularly over
the big head, thick neck, broad shoulders, and powerful front legs. Behind
the shoulders, where a full-grown bison is about six feet tall, the back
slopes downward to the relatively small hindquarters and tail. Despite the
terminal brush of hair on the tail, these parts of a bison seem to belong
to a less imposing beast, for they are thinly clad. In winter the animal
shields them from the cold wind either by facing into it or by facing
outward as one member of a ring with tails together, grouped for mutual
protection.

While plodding along, a bison seldom raises its head far. Its goatee,
longer in males and shorter in females, almost touches the short grass. The
neck slopes upward to the shoulders, giving the animal a dejected mien.

The front legs bear most of the weight, with the head and neck counter-balancing the more posterior parts. A boisterous calf will lift its mother's hind feet right off the ground if it butts her body from below where her teats are located. The calf has excellent leverage when it kneels below its mother to reach her milk.

Most calves are born in May, after about nine months of gestation. To bear her single calf, or a rare pair of twins, the mother walks a short distance from the herd among which she has been feeding. At that season, with the grass growing so rapidly, the bison graze with unaccustomed deliberation. Actually, they are more alert than usual for the approach of four-legged predators. They rally quickly to the defense of any mother with a newborn calf and hold the meat-eater at bay until the youngster can get to its feet. By forming a circle around the calves, the adults give them warmth, shelter, and access to nourishment while presenting an almost solid wall against attack.

Until men with guns deliberately destroyed the vast herds of bison on the Great Plains, both to make the land available for settlers and to destroy the resource upon which the Plains Indians maintained their independence, the behavior of the bison and of the wolves that followed each herd interacted to the benefit of the prairie grasses. Whenever a nomadic bison herd increased until the supply of grass failed to satisfy the hunger of every individual, the bison increased their rate of travel. Old animals, some in their twenties, and any with infirmities tended to lag behind, where they no longer had the herd to protect them from wolves. If the faster pace did not lead to improved nutrition, the herd divided and then might travel more slowly, grazing with increased efficiency.

Wherever the food for bison proved to be totally inadequate, the herd splintered even more. Groups composed of fewer than a dozen adults separated off with their calves. But so small a group could not form an effective ring when wolves attacked. Instead, the adults panicked and stampeded. As the mature bison galloped ahead, the predators picked off their young, then killed each adult as it tired, and finally the last one. Soon the bison population shrank to match the depleted supply of grass.

Wherever the bison roamed, wolves were always nearby, waiting for an opportunity. Yet they rarely risked approaching any healthy animal in a herd, even if it was off guard. A bison can move quickly when it wants to, and is a formidable opponent at close range. The wolves and other predators watched for accidents from which they could benefit, and kept the edge off their hunger by pouncing on smaller prey or eating less nutritious meals of vegetation.

No wild predator offers much of a threat to a healthy bison while it is a member of a successful herd. Most herds, in fact, center around an old cow who has learned much that is useful in her twenty years or more. Accompanying her are calves of several generations, of both sexes, but no young bull that is ready to challenge the big old bull roving in the vicinity.

Nor are small groups of full-grown bachelors to be attacked with impunity. They can charge in a devastating counterattack, or gallop off to join a herd and benefit from community action.

These family habits of the bison persist today on the National Bison Range near Moiese, Montana, where a fixed population of 400 is maintained; in Yellowstone National Park, where 800 have a safe place to live; in Wind Cave National Monument, South Dakota, which accommodates 370; in the Wichita Mountains National Wildlife Refuge, where 900 find food and space; and in Custer State Park, South Dakota, which is the largest state park in the country and provides sanctuary for 1,300 bison —the biggest population of these animals anywhere in the United States. Canada offers hospitality to somewhat larger numbers in a special Wood Buffalo National Park, which extends from the northeast corner of Alberta to the south shore of Great Slave Lake.

We have accompanied the manager of the National Bison Range on his regular rounds of the sanctuary hills. In a jeep, which excites far less antagonism in the herds than any party on horseback, we felt far safer than afoot where bison-proof and climbable trees were so few. The big animals generally ignored us. They continued with their grazing or their nibbling on shrubbery, their siestas or their dustbaths, their ponderous scratching of elusive itches. It was August, which is the least suitable time of year for any close approach. We had to rely upon binoculars and telephoto lenses, particularly when the one ivory-white bull turned in our direction. Apparently his gray-blue eyes gave him clear vision. He was the dominant animal in his herd, and seemed well aware that by moving toward us at any time he could send us on our way.

This magnificent white bison, called Big Medicine, climaxes our motion picture records. Yet they recall for us the touch of sadness with which the manager, John E. Schwartz, spoke of the rarity of white bisons. He knew of only two others in recent years. One of them, blind since birth, lived in the National Zoological Park, Washington, D.C. The other had been sighted in a far-northern herd. Big Medicine eventually died of natural causes and became a permanent exibit in the natural history museum in Helena, Montana.

We asked the range manager about the rate at which the herd increased in numbers. We remembered that in the 70 years following 1830, wholesale slaughter of bison reduced a population of more than 40 million to 541 in 1889. Now the descendants of those survivors had had 70 years of complete protection during which to restore the bison population within the limits of the food available inside the sanctuary boundaries.

The early life of each bison is beset with hazards. Not until each youngster is about three years old has it passed the danger zone. Mortality is greatest at birth or just before, and continues to be high for the first year. Young bison often die during the cold months of their first winter even where custodians provide food supplements whenever the natural sources shrink to a low level. No one is sure whether this loss has always been customary or if it is a consequence of confinement, now that bison can no longer follow their inborn pattern of seasonal migration.

The surviving herds on the reserve lands encounter fewer predators than in the past, because the wolves have been eliminated (except in Canada's Wood Buffalo Park) and, in most areas, the grizzly bears as well. With smaller herds, the sharing of diseases and parasites is less likely. For the bison in Yellowstone National Park, diseases, parasites, and predators all seem unimportant. Yet the herd shows only a slow natural increase.

The Eurasian bison *(Bison bonasus),* known as the wisent, has not yet reached the population size at which no more can be given space and food. Centuries ago, it was exterminated in Asia. The survivors became forest dwellers, with a few herds in the remote parts of the Caucasus Mountains and somewhat more in eastern Poland. A token group received some protection in both areas until World War II, particularly in the nationalized estate called the Bialowieza Woods near the Russian border. All of these wisent died of gunshot wounds before the end of the war, when weapons and ammunition were easier to get than meat.

Fortunately, about 60 wisent remained alive in various zoos. Scientists who longed to restore at least the Bialowieza herd to its former condition made special efforts to obtain a breeding stock from among these captive animals. In many instances, the zoo dweller had to be rejected because its history showed that it was a hybrid—bred in captivity from an experimental cross between an American bison *(Bison bison)* and a European wisent *(B. bonasus).* Although the hybrid might be fertile, it could be used only as a last resort.

We wonder how much the family life of the reconstituted herd of wisent in the Bialowieza forest today resembles the pattern of centuries ago. For zoo animals, with more opportunities to interact with people than with other members of their own kind, the inherited behaviors must be strong indeed to produce a semblance of wild activity in a large and wooded sanctuary. Amazingly, the Bialowieza bison have recovered their habit of mating in August. They slip through the forest in herds dominated by an old cow while an old bull—the victor in every fight—patrols the perimeter and drives off contenders for his role. Surrounded by a natural environment, the wisent seem to have recaptured their old ways, as if an opportunity was all they needed.

The Polar Oxen

We can certainly agree with Henry Kelsey, the Canadian explorer who discovered musk oxen *(Ovibos moschatus)* in 1689, that these animals of the Far North resemble little bison that have let their hair grow long. A full-grown bull musk ox weighs 500 to 900 pounds, and stands five feet tall at the shoulder, where his hairy coat is curly. Elsewhere it hangs straight, over his face, hiding his short neck, his front and hind legs, and his four-inch tail. Generally the coat is ankle-length and dark brown or black. It contrasts with white hairs on the exposed parts of the feet and with a pale saddle-shaped marking on the back.

Recent discoveries prove that a musk ox neither has musk nor is an ox. Musk is the secretion of an abdominal gland on the musk deer, whereas the scent of a musk ox is an odorous product of a modified tear gland. A musk ox exudes the material when frightened, and spreads its odor by wiping its eyes against the hair on its front legs. The musky smell is sometimes recognizable from a hundred yards away. The chemistry of musk ox blood supports the claims of paleontologists that these animals evolved from the goat-antelopes rather than from oxen. The nearest living kin of musk oxen are the Rocky Mountain goat *(Oreamnos americanus)* and the European chamois *(Rupicapra rupicapra).*

Like bison, the musk oxen protect themselves by forming a close ring with any calves at the center. All of the adults stand with their horned heads pointing out. But musk oxen are not content to wait on the polar tundra until a predator goes away. Instead, one or more bulls at a time rush from the circle to attack. Extraordinarily nimble, they attempt to gore or trample on any wolf, dog, or person they can isolate. Then backward they go into the group formation, to glare and watch for another opportunity to divide and destroy their tormentors.

Through the dark days and long nights of the arctic winter, the musk oxen rarely separate into groups of fewer than 20. As many as 100 may socialize and paw through the thin snow for vegetation on the rocky tundra. The weather is often scarcely better in April and early May, but the herding behavior grows less intense. A herd of two dozen becomes more usual. In it, about four animals will be pregnant cows. Each in turn, as her hour for giving birth arrives, goes a few steps from the others—no more. Seldom does she lie down. Instead, she lowers her head almost to the ground, with fore and hind feet close together, and contracts the muscles that expel her calf. Shivering, the newborn animal frees itself from her birth canal, slides to the frozen ground, and struggles immediately to stand up. For as long as she will stand still, the calf leans against its mother, sharing her heat and letting her shield it from the wind. Its woolly covering gives it almost no protection from the cold at first. The mother, meanwhile, tries to appease her hunger, which is rarely satisfied all winter long. As she moves along, the calf must keep up. If it falls behind during the first hour, before its fur is dry, it may easily freeze to death.

Musk oxen have survived despite adversities that brought disaster to many an early polar expedition. Their adaptations give them a slight margin of safety in a precarious land. While a well-fed cow musk ox is capable of achieving maturity and bearing her first calf at two years of age, and another annually for many years, on the high Arctic tundra her performance falls far short. A cow is likely to be four or five years old before she has her first offspring, and probably won't complete another pregnancy for two or three more years. Like all musk ox mothers, she will nurse her calf for three months and wean it in midsummer. She needs the rest of the short growing season to restore her own reserves. The calf is on its own, protected only by the social reactions of the herd among which it remains. About half of the calves live to be a year old. The rest perish, benefitting the wolves.

So far, an artificially introduced herd of musk oxen has fared well on Nunivak Island in the Bering Sea, not far from the mouth of the Yukon River. Almost 500 of them live there today, all descended from 31 calves brought from Greenland in 1935 and 1936. For a while, the bleak island seemed a poor home. The alien herd fluctuated in numbers, rising and falling and rising again. In 1947, the total reached 47. As though number 48 brought luck to the herd, the population began to increase fairly steadily thereafter. About 7 musk oxen die for each 16 calves born. The chief killer proves to be harsh weather rather than predators or disease. Now the herds on Nunivak have reached a critical size, for they are fully utilizing the winter range available to them.

Fortunately, musk oxen have an energetic friend. Mr. John Teal, an arctic ecologist, dreams of domesticating these polar animals and accustoming them to life on his farm in Vermont. He is convinced that the musk ox is adaptable and would contribute to human welfare both palatable meat and a wool that is superior to the best cashmere. The wool comes from the dense undercoat of the musk ox, and need not be combed out as herders must do with Cashmere goats. Each spring a musk ox sheds about six pounds of the material, compared to three ounces from a goat. No simple way has ever been found to collect the wool, known by its Eskimo name of qiviut, from the low shrubs of the tundra barrens. But gathering it from captive musk oxen should be easy.

Since Nunivak Island has calves to spare, Teal needed a suitable method for capturing them. It could not be so crude as the technique used for capturing the original breeding stock in Greenland for Nunivak; then it had seemed necessary to shoot almost all of the adult musk oxen that were so valiantly defending their young. First, suitable facilities for receiving and caring for musk ox calves were arranged at the University of Alaska in Fairbanks. Then Teal and his assistants tried to drive a few adults and their calves into the cold waters of a bay along the shore of Nunivak Island, so as to be able to lasso the calves from a canoe. Presumably the parents could not interfere, since they would be swimming for their lives and unable to form a defensive circle in the water.

The musk oxen on Nunivak proved to be much too elusive and belligerent to be driven into bays. A musk ox can swim well when it wants to, but can't be forced into the water. Another technique was tried: dispersing a herd with the downdraft from a small helicopter, then capturing the cowering calves before the adults could reassemble. This method worked so well that a skilled pilot and the men on the ground were soon capturing six calves a day, without physically harming one of them or any of the adult musk oxen. Teal could even choose how many young bulls he wanted and how many heifers. Excess bull calves were freed again.

At Fairbanks, the horn buds on the young musk oxen were removed by the same procedure used so routinely with domestic cattle. To everyone's surprise, with the possible exception of Teal's, the wild calves accepted their substitute diet and new surroundings in just a few days and began to grow at a prodigious pace. They became tractable, friendly animals, perfectly willing to cooperate with man's first generous gestures toward their kind. Hopes soared for developing another herdable mammal in the ranch country of the North, one that might eventually help the precarious Eskimo economy.

The Elephants Among Seals

In the Southern Hemisphere, August brings an equivalent to February north of the tropics. We hardly expect it to be an amorous season for any mammal swimming near Antarctica. Yet the giant among seals—the southern elephant seal *(Mirounga leonina)*—arrives then around the shores of remote islands and hauls up on the rocky beach. The largest of these colonies, with about 300,000 elephant seals, is on South Georgia. A third as many come to Macquarie, and an equal number to Kerguelen. Smaller colonies are growing on bleak shores of the Falkland Islands, of Crozet, Campbell, St. Paul, and Amsterdam, and even on Tristan da Cunha in the middle of the South Atlantic Ocean. Altogether the world now has about 800,000 of these monsters, although 70 years ago they seemed headed for extinction.

The biggest bulls are the first to come ashore. Initially they seem reluctant to leave the icy water. This seems doubly understandable, for hauling a four-ton body over the rocks requires tremendous effort and, until the animal can return to the buoyant sea, he will get nothing to eat or drink. It will be late October—the equivalent of April in the Northern Hemisphere—before his parental duties will be fulfilled and he can return to the sea.

For the first three weeks of August, his large brown eyes may not even see a cow of his species. The pregnant ones swim close to the islands but carefully pass by any place where they can see a huge bull contesting for

a breeding platform. They seek instead much quieter places to haul out along the shore and give birth to their 100-pound pups. For three weeks each mother devotes herself to the welfare of her single youngster and then abandons it. On her rich milk, which is about 80 percent butterfat, the pup can grow to weigh 400 pounds in this short time. It will measure a full four feet from the tip of its nose to the ends of its outstretched hind feet. Already the thick woolly coat of black hair it was born with is being shed from its back and belly. Exposed by the molt is a pale gray covering of almost silky hairs. Later they will stiffen and thicken. By the time the pup is a year old, they will form the sparse bristly covering that is characteristic of all elephant seals.

For the next few weeks, the giant seals of both sexes and all ages act like animated chess men. They move about on the shore according to inherited rules in a pattern that has brought success to the species over millenia. The most spectacular performance is that of the big bulls, which have been battling furiously all day long while the cows have been attending to their pups. The centers of contention are called breeding platforms, each a broad, shallow depression in the rocky beach, big enough for a bull and a harem of from 12 to 30 cows. Rarely is a bull sufficiently powerful to seize and defend a breeding platform before he is ten years old. Only with good fortune can his vigor continue year after year and let him retain this eminent status. Every other bull wants to usurp his site and his prerogatives. To hold on, he must win every battle.

As in few other kinds of life, the bull elephant seal never seems to accept whatever status he has achieved until he has a breeding platform to defend. He must challenge every other bull he meets and engage in vicious combat until one of the two backs away or escapes into the water. Each confrontation follows the same pattern. The two males approach each other, rearing up as tall as possible on their front flippers. Both animals exhale great snorts of air, which inflate the bulbous enlargement of the nasal area for which the elephant seal is named. Apparently this is a resonator, for the sound of air entering and escaping from it becomes a bellow that can be heard for a mile or more.

Once within slashing distance of the competitor, a bull sea elephant sweeps his head downward, raking his opponent's hide with tusklike canine teeth. Thick hide over the neck region gives each animal fair protection, but soon the hide is bleeding, later to become rough and scarred. The special target is the vulnerable nose. More than a foot in length and about eight inches wide, it swings up out of the way as the elephant seal bites at another male.

As soon as a cow weans her pup by simple abandonment, she returns to the water and propels herself more easily toward a breeding platform. Dodging any nonharem bulls that are patrolling the coast, she somehow chooses a giant mate. She hauls herself ashore and wriggles to his breeding

Southern elephant seal (*Mirounga leonina*)

platform, staying out of his way if he is in the midst of a battle with a neighbor. As soon as he can, he accepts her and climbs over her to copulate. How she survives his enormous weight is far from clear, for she almost disappears under his body. His attentions, however, are brief, if only because adjacent bulls give him no peace. As though in compensation, he will return to the same new cow several times on the first day she is in his harem, and again on the next if she will receive him. By then she may have left, and her place taken by another. Once pregnant again, a cow is anxious to satisfy her hunger. She swims out to sea where she can find the fishes and squids that are her basic diet.

The deserted pup crawls higher on the beach and joins others of the same age—generally behind the breeding platforms and far from the water. There the pup plays and sleeps and continues its molt for a week or two. Then, avoiding its elders, it progresses to the edge of the sea as though needing to have wet rocks to scratch against in getting rid of the last of its black woolly natal coat. The pup is still a big baby, with no need to eat until it is about seven weeks old. It lives on the fat that it accumulated before it was weaned, and only later feels a need to teach itself how to hunt. Its first meals are of crustaceans, which are slower than the prey of the adult seals.

For a while, the margins of the breeding islands seethe with fat bodies. Some pregnant females may still be arriving. Mothers that have left their pups are moving toward breeding platforms. Young bulls that have not yet been able to win a place on land are battling with one another. Young females that have never mated before come into the shallows and distract the young bulls, which find a few moments to copulate before returning to the fray. Among all this confusion, the pups that are completing their molt stay out of the way as best they can. All of these preparations culminate as the elephant seals satisfy their various social needs and slip off into the sea to swim away.

No one is sure how closely these amazing animals follow the edge of the pack ice toward the shores of Antarctica as the sun's warmth opens the way. The progressively shorter nights and higher sun let the elephant seals find prey in fresh abundance at higher latitudes and greater depths. After the solstice in December the web of life in the waters of the Southern Hemisphere attains its greatest richness. It sustains seals of all ages as they grow and add to the thick layer of fat beneath their skins.

The elephant seals respond to the changing proportions of night and day toward equinox in March (the equivalent of September in the Northern Hemisphere) by coming ashore again. This time their fast is relatively brief, and battling is at a minimum. They come ashore because their summer coat is falling away and a new winter coat is growing out. As if this did not itch enough, the animals are plagued by biting insects and an uncomfortably hot sun. Each elephant seal digs itself in on the beach by wriggling like a fat maggot and tossing sand or pebbles over its back with its front flippers. Much of the summer covering slides off each big body as the elephant seal sleeps restlessly. Piles of debris accumulate between the seals, as though they had taken refuge in the craters of a battleground.

Formerly this molting season in the autumn of the Southern Hemisphere brought the elephant seals their greatest danger. Men who might profit from killing the animals for meat and oil were most likely to arrive by ship in March. They came regularly during the nineteenth century until they could find too few surviving elephant seals and other marine mammals on the shore to make their trips worthwhile. Subsequent neglect and then disdain for harvesting this resource, because of low economic value of the meat, almost unsaleable pelts and fur, and meager yield in oil as compared to whales, allowed the animals to rebuild their populations.

As we consider the schedules by which the elephant seals, the musk oxen, and the bison start new lives each August toward maintaining their populations in their separate worlds, it helps to look at a globe of our planet. It reminds us that the elephant seals and the bison are actually animals of the temperate zones, whereas the musk oxen are truly polar. The South Temperate Zone is mostly water—an ideal realm for elephant seals to grow huge in. The tips of the continents and the various islands are far apart, surrounded by stormy seas. Yet the length of day and the slant of the sun varies no more than in the Northern Hemisphere between Stockholm and Barcelona or between Whitehorse in the Yukon and Denver, Colorado. The elephant seals that occasionally wander to the shores of St. Helena in the South Atlantic have invaded the tropics; they are as near the equator as Barbados or Rangoon. The home base of the elephant seals seems far away and cold and bleak.

Long ago, elephant seals swam through cold waters along the eastern edge of the Pacific Ocean and accommodated their lives to existence in the Northern Hemisphere. They became northern elephant seals *(Mirounga angustirostris)* instead of southern ones *(M. leonina)*. They attained a smaller size, ranging in summer as far north as the islands of southeastern Alaska, and breeding on rocky islets from Baja California to San Francisco. There too they were exploited by adventurous fishermen until the population all but disappeared. Today they have recovered somewhat, and about 10,000 northern elephant seals haul out annually to fight, to pup and nurse, and to mate in a northern equivalent of August—from mid-December to mid-March.

Scientists from the University of California have been studying the family life of northern elephant seals since the early 1960s. Unlike their southern relatives, these seals establish no fixed breeding platforms. The dominant bulls move about on shore, battling the less dominant males for a position central to a group of cows. If the females shift their location, the big bull follows them. This arrangement works well too, with 4 percent of the males mating with 85 percent of the females. If the same bulls gain top status year after year, and if they actually fertilize the eggs of the cows with which they mate, the population of elephant seals must be remarkably inbred.

The dominant bulls have a strong influence over the breeding colony in still another way. Each rookery of northern elephant seals has its own special dialect. It shows in the threat calls which the bulls bellow forth to warn away an encroaching neighbor or to dare him to do battle. On one island, each vociferous male will rear back and emit from five to fifteen pulses of loud sound at one-and-a-half- to two-second intervals, with all pulses and intervals essentially alike. On another island the number of pulses may be slightly greater and the final one may be prolonged in a great roar. A third island dialect may include in each threat call only about half a dozen pulses, each three times longer than those of other dialects and well spaced as though for maximum effect. Since elephant seals swim from one island to another and have recently increased their breeding range in this way, inherited differences can scarcely be credited with causing the dialects in threat calls. More likely the aspiring younger bulls imitate the sounds of the dominant individual. Whether this helps them to share the favors of cows in his harem has not been proved. Since the same dialect reappears on the same island year after year, the distinctive features may assist each pregnant female to find again the harem master whose pup she is ready to bear.

CHAPTER SEVEN

Signals of September

ANYONE WHO HAS ever heard the mating call of the bull elk, the bull moose, or the red deer stag in the Old World, would understand why each September first special orders go out from the head offices of Canada's major railroad. They direct the engineers of all diesel trains crossing the western provinces to use the electric horn as little as possible until after mid-October. The blaring signal might be mistaken by an elk or moose, causing it to come charging out of the brush beside the track and collide fatally with the train. To an English gentleman, the error would be understandable because the prolonged note of the hunting horn, intended to induce antlered targets to show themselves, is a September sound on many a big estate.

The Red Deer

In the Old World, only the stags—the males—of the red deer *(Cervus elephus)* would be the coveted trophy animals, not the hinds—the females —which are smaller and normally lack the magnificent antlers that adorn stags all through the rutting season. Stags ignore the hinds until September, while the hinds have been living separately with their calves in little herds. Then, for each male in turn, the rising level of sex hormone in his

bloodstream triggers a change in behavior. He separates from the other stags, with whom he has been associating amicably and silently, and trots over to the nearest herd of hinds.

At first, the male is all display. He bellows forth a deep *raaah,* known as a roar, and repeats this call endlessly at intervals of less than a minute. Brushing his antlers against shrubbery, he tears it up and makes "rutting places" by scratching away almost every trace of vegetation. Even to a human nose these rutting places have a distinctive odor. Most of it comes from secretions that the stag rubs off from special glands on his feet and beside his eyes. The appearance of the agitated male changes too: his neck enlarges, making his mane more noticeable. And he wallows in every mudhole he can find, or in his rutting places, getting his coat covered with dirt.

As more stags become excited, they confront the first comers. Running around a herd of hinds and calves, they meet each other, spar with their antlers or simply stand and roar. For at least a week the stags get little sleep, although they lie down for a few minutes occasionally, as though exhausted. All day and all night their confrontations continue. Often the peak of roaring comes around midnight. Gradually the herds of placid hinds get divided up into harems.

The hinds remain amazingly oblivious to all this commotion among the stags. Certainly these females are not deaf. Nor can they fail to see and smell the rutting places, if not the stags themselves. Probably each hind reacts by becoming increasingly stimulated and ready to mate. She will communicate this state to the stags by secreting odorous substances of her own as she walks sedately about the woodland.

Until this signal comes, each stag performs as a would-be harem master. He dodges around the hinds, blocking their way in some directions, nudging them in others, directing them if possible toward other females of his kind. The hinds are docile enough, although sometimes reluctant to move in the direction the stag wants. His threat of a kick or two becomes a convincing argument. Rarely does a stag lower his antlers to propel a hind where he chooses for her to go. He saves this gesture to threaten any other stag that approaches his harem. If the intruder has a smaller rack of antlers, he usually recognizes the inequity and turns away promptly. If the newcomer is a bigger animal, the harem master abandons the hinds; the bigger stag gains them without a fight. It is only when two stags are almost evenly matched that a battle is required to determine which is the stronger, more virile, or more determined. The clash must not last long, however, for the hinds might wander off or be herded away by other stags, leaving neither contestant a reward for winning.

All of the activity in rutting season centers around the hinds. They try to continue their preferred schedule—moving about to graze and browse during the morning and late afternoon, and lying down for a noonday

Red deer (*Cervus elephus*)

siesta and an all-night sleep. Even when a hind shows by her scent that she is ready for the stag, she receives him only as she walks along. He may mount her a dozen times in succession before he is satisfied. Most hinds reach this condition before the end of September and become pregnant at once. A few are slower and can accept a stag's attention during the winter months, in some instances as late as March of the following year.

The stags cannot cooperate for so long a time. As their rutting period nears its end, they reserve most of their roaring and their conflicts for midday. By mid-October they are silent again, although still willing to mate with any receptive hind until December. Toward winter solstice, the lengthening nights dampen a stag's excitement. In January the lessened concentration of male sex hormone circulating in his blood allows natural processes to weaken the anchorage of his antlers upon his skull. These weapons and symbols of sex and status drop off stags that are in their prime by February. Old stags lose them by March, and the one-year-olds that are following their mothers often don't lose them until May.

After the rutting season, the adult males herd separately in little social groups, generally accompanying an old hind who has no calf and for some reason shuns others of her sex. One might think of her as a sort of housekeeper, except no one can see that she does anything to make life easier for the stags. Most of the hinds join larger groups that provide protection for calves as they grow toward maturity. The distinguished behaviorist Dr. Frank Fraser Darling, who has studied red deer most extensively in all seasons, regards their herding patterns as those of a matriarchal society, in which the hierarchy among stags according to antler size and vigor may be a secondary phenomenon.

The selective force that determines how many and which red deer will survive is the availability of food, particularly after winter storms. The full-grown hinds have a double advantage, for they can rear up highest into the lower branches of trees to browse when nourishment lower down becomes scarce, and they also have the greatest reserves of fat from good eating during the preceding seasons, which they share with the unborn calves developing inside their bodies. The stags enter the winter with little fat, for they have put eating in second place through the rutting season and have difficulty after mid-October catching up on meals in the temperate woodland. The growing calves are the most vulnerable, for their nutritional needs are greatest and their ability to reach tree branches often inadequate. They starve first wherever too many red deer are living in a limited area, or if a hard crust of shallow snow and severe frost locks away the plant foods that can be pawed for on the ground.

Conditions improve toward the end of May or in early June, when the pregnant hinds bring forth their calves. Usually each female bears just one. Twins are rare. Except for its white spots, the calf has the same reddish brown color as its mother's new summer coat. Like all red deer, it wears a bright rusty rump patch or even a yellowish one. Until autumn, when it is weaned, it will stay as close as possible to its mother, learning from her and benefitting from her protection against predators.

The Wapiti

Few of the settlers who came from Britain to the New World in the 1600s were landed gentry who had experience hunting stags with horn and hounds. Poachers familiar with the big estates may have been among the early colonists, however. Many had seen red deer and were aware that a much larger animal of the deer family, not found in the United Kingdom, could be found in Scandanavia. Probably none of the colonists had encountered this giant creature of northern Europe, but they knew its name. It was an elk.

In the deciduous woodlands of New England, in the Berkshire hills, and along the Appalachians, the colonists encountered a giant deer whose range included the northern two-thirds of North America. Full-grown males weighed about 750 pounds, whereas the biggest red deer stags would rarely exceed 500. The Indians called the animal a wapiti. If the colonists had adopted this word, they could have avoided much confusion later. But they decided to call the big American deer an elk, and the name stuck.

Biologists use scientific names to keep these handsome animals sorted out: *Cervus elephus,* the red deer of Britain and woodlands of the Eurasian continent; *C. canadensis,* the "elk" or wapiti of American woodlands; and *Alces alces,* the elk of Eurasia and the moose of North America. No one explains why the male and female of the red deer are stag (or hart) and hind, whereas those of wapiti and moose are bull and cow. The young of all are calves.

Today the wapiti have been killed off from all eastern portions of their former range. They are often called Rocky Mountain elk because their woodlands there remain less spoiled and their populations continue to maintain a semblance of the past. Our first encounters with wild elk came in western Wyoming and Montana, where the bachelor herds of bulls trot back and forth across the mountain meadows and the cows with their calves conceal themselves by day among the willows and alders in the wet valleys. Under cover of darkness and into the early morning, they emerge into the open to find the plants they prefer to eat.

Like the red deer stag, the bull wapiti keeps silent except in rutting season, when he bugles with nose elevated and flexible lips forming a large O. The signal starts low in pitch, rises as it grows louder, and falls rapidly to end in a grunt. Early in September, the bugler is likely to be a giant wapiti, because those with the largest arrays of antlers generally attain sexual readiness earlier than those with smaller ones. The biggest bulls press forward faster than their normal walking gait. They trot tirelessly to find a herd of cows and calves that will accept their company. The only wapiti that are excluded at this season are the young bulls in their second summer, when their antlers are mere spikes. They have been living by themselves ever since their mothers rejected them and drove them from the nursing herd, just before the new calves of the year were born.

At intervals during the day, and often while the cows are resting in the shade, the bull digs a shallow pit by pawing at the earth with first one front foot and then the other. He may lower his head and use the front tines on his antlers to help in this excavation. He urinates into the pit, scrapes the mud around, urinates on it some more, then lies down and wallows—bugling—in his own excretion. We cannot doubt that this wallowing adds to the bull's aroma. Whether it makes him more attractive to the cows is another matter.

Each cow maintains her sociability with other cows. She nurses her own calf and tolerates theirs while her reproductive system quietly develops on its own schedule toward a day when she can be impregnated. Her only communications seem for her calf, gestures and little coughing barks to which the calf responds with soft sounds of its own. But when her day for mating arrives, her scent lets the bull know without mistake. He singles her out and follows until she is close to the periphery of the nursery herd. There he performs his distinctive courtship. He rises as high as he can on his forelegs, crouching a little on his hind legs as though getting ready to sit. Lowering his nose, he elevates his antlers in a most impressive fashion, aiming his display directly at his chosen cow. If she does not give the countersign, he walks along awhile beside her, half facing her, and tries again when she stops. Often this posturing goes to waste in the early morning hours, and he must repeat it in the late afternoon if he is to get his way.

No one credits the cow wapiti with being coy. She simply cannot hurry the events inside her body. When she is ready, and not before, she will let him lick her rump and her perineal region, which gets him even more excited. She walks forward slowly with her head low, often with her neck turned so that one eye looks toward the sky and the other toward the earth. If he does not mount her then, it is because he has been disturbed in some way or is busy repelling a rival. The cow soon abandons her submissive posture. Within hours she loses her receptiveness, not to regain it until about eighteen days later.

Dr. Valerius Geist of the University of Calgary, Alberta, tells us that he can distinguish three different kinds of threat by which one bull wapiti attempts to drive off another without actually engaging him in battle. One is a threat to kick or bite, raising the head high and laying back the ears, and either elevating one front foot in readiness or raising the upper lip until the small, sharp, canine teeth show clearly. These gestures are particularly common when bulls quarrel before their antlers harden. A bull with good antlers is more likely to lower his head while facing his antagonist, as though to show off how many sharp points he might jab against the other. Sometimes the competitor does likewise, and the antlers touch for a moment of sparring. Toward a lesser bull, a big wapiti is more tolerant and merely turns broadside, perhaps looking away to let the smaller animal see how much bigger and more formidable the intimidator's antlers are even from behind. If any sound is made during these displays, it is a short bark as though to demand attention.

The pregnant cow wapiti, toward the end of her 250- to 262-day pregnancy, uses threat gestures with her front feet to drive off the yearling calf that has been accompanying her. She will not tolerate any interference while she is giving birth, or until she no longer feels a need to keep the new calf hidden and can lead it forth to meet the rest of the nursery herd. Young mothers, who have had only a year or two to experience the conflicting pulls of maternity, sometimes continue to protect their latest calf in September by avoiding any bull wapiti even when their reproductive systems are ready to receive him. By October, the nervous cow is more likely to let the bull approach. Her calf is bigger and more able to look after itself, and her inherited schedule is calling for her to be sociable with any wapiti bull. Soon all these animals will be mingling freely in mixed herds. Food, not sex, will become the primal drive.

In America's western mountains, the wapiti migrate up slope and down according to the season. In late autumn they descend and try to survive on grasses and sedges, although these plants give them only marginal nourishment. Many wapiti are so weak by spring that they die while attempting to follow the line of melting snow up the mountains. Coyotes and other predators and scavengers feast on the unfortunates. The remaining wapiti drag themselves to the high meadows where fresh herbs are bursting from the wet soil and the low shrubs are sprouting new growth. Willows and lupines act like a tonic on the wapiti. In just a few weeks, their signs of malnourishment disappear. Their coats and eyes gain luster. The bulls are growing new antlers and moving off into bachelor herds. The calves put on weight rapidly. The pregnant cows begin to show their condition. Once again these giant deer have won in their contest with the winter. They are ready to restore their population just as they have always done.

Today, little herds of sociable wapiti are being introduced into some parts of America where they were extirpated a century ago. Once again it is possible to encounter these handsome animals, which attain the size of riding horses, east of the Great Plains. Hikers can glimpse a bachelor herd trotting swiftly across a meadow, and admire the magnificent antlers with points upturned above the animals' backs. A shed antler becomes a treasure for some person to discover. The restoration of the wapiti adds little competition for other kinds of deer because each species chooses its own diet in every season and location. Wapiti prefer plants that domestic animals generally ignore. Moreover, despite their formidable antlers, these handsome mammals offer no dangers to humankind. In no instance that can be verified has a bull or cow wapiti attacked a person in an unfenced area from which a silent escape was possible. Wapiti shy away at the scent of people even in September, when a bugling bull will normally hurry over to investigate any imitation of his signal.

The Unpredictable Moose

Meeting a moose in the wild provides a far more hazardous experience. A bull moose is as likely to charge as to wheel and run away. A cow generally has a calf close by and is belligerently protective if she senses that it is threatened in any way. Often the cow and her calf have a bull in attendance. We prefer to keep a pond or stream between us and any adult moose, and to enjoy our observations through binoculars or a telephoto lens.

A full-grown bull moose is the largest member of the deer family in the world. From the end of his rounded nose to the base of his tail, he may measure more than nine feet long. At the shoulders he will be more than six feet tall. His three-inch tail will be much less conspicuous than the fleshy, hairy "bell" that dangles underneath his throat. Even his dark color —almost black in winter and more reddish in summer—makes him loom especially big as he sniffs the air and eyes his surroundings from a height of about seven feet. Above his head spread massive antlers perhaps six feet across, with broad thick blades and an array of upturned tines.

The largest moose are those of the Kenai Peninsula in Alaska, where a special reserve has been set aside for their safety. Measurements prove that the Eurasian animals ("European elk") are distinctly smaller, although those we observed among the shrubbery around ponds darkened by tall spruce in Swedish forests seemed huge. We found them by stopping to explore on foot wherever we saw a bright yellow warning sign with a black silhouette of a bull moose head along the highway.

We waded through the soggy peat between the shrubs, eagerly and apprehensively alert for any sign of moose. Yet we could not help noticing that this same community of vegetation, dominated by spruce, fir, and peat moss, had occupied almost identical areas during the interglacial periods of the Ice Age while humankind was evolving farther south. Then the elk of northern Europe, and of Scotland and Ireland too, had been the one known as the Irish elk *(Megaceros hibernicus)*. The tine tips of its antlers fringed up beyond the blades to a record width of eleven feet from side to side. So far as we can learn, it was extinct before human hunters reached the north country where they might have met it face to face. Skulls with antlers still attached have been removed from natural storage in ancient peat bogs.

The fossilized remains of the Irish elk reveal nothing of the family life the immense beast enjoyed. They focus our attention, instead, on discovering why members of the deer family produce antlers. Surely the answer should be most obvious among those that possess the largest. One suggestion appeared in 1968 in the respected pages of *Nature,* the weekly journal of the British Association for the Advancement of Science. It pointed to the heat that a warm antler must dispel while it is still covered by skin and circulating blood. Perhaps an antler is merely the structural support for a cooling system, and remains in the autumn after its summer need is past. Yet the skin (the "velvet") ordinarily dies and falls away before summer's heat is gone. And the biggest antlers are found on animals in higher latitudes and elevations, rather than on lowlands near the Equator where cooling would presumably be most important. Moreover, on each animal, the antlers increase in dimensions either much faster than the production of heat (as measured by the weight of warm body tissues) or considerably more slowly (in the roe deer of Europe). As the body of a moose doubles in size, the weight of its antlers may leap ahead four- or fivefold.

Dr. Geist believes the rewards for producing antlers more likely lie entirely in social values. They benefit a bull as he establishes his rank in his own community and maintains it through his prime. Antlers help him impress other bulls. He can use them to ward off attacks by other members of his kind and to fight when this becomes necessary. They are symbols as much as guards and weapons. To this extent they are products of evolution parallel to the tusks of elephants and the strange nose horns of rhinoceroses.

The bull moose grows his new antlers between May and August and exposes them by scraping and rubbing off the velvet against favorite trees. A fir four inches in diameter and lacking lower branches is eminently suitable. As its rough bark becomes smeared with bits of skin and clots of blood, the antlers shine bone-white with bloody streaks in the grooves where vessels formerly pulsed. Soon, however, resin, dirt, and the juices of other plants stain the antlers a rich brown. The moose hastens this transformation by thrashing the shrubbery as though it were an opponent. Sometimes he kneels down to reach blueberry bushes and tears them from the ground if they can't resist his slashing antlers.

Kneeling comes naturally to moose of both sexes. Their legs seem too long to be kept straight, in a standing position, while the animal grazes on the low sedges and horsetails that have special appeal. Standing upright is fine for higher bushes, for the needles on drooping branches of balsam fir, or for the branch tips and foliage of fast-growing deciduous trees such as aspen, birch, mountain ash, and willow. Long legs are ideal for stepping over fallen trunks in a swampy area, or for wading through bogs and mudholes. They can also be used efficiently for swimming.

One of our favorite vantage points for moose-watching is a bank overlooking some beaver ponds at Moose, Wyoming. There we have observed and photographed a cow moose standing shoulder deep in the water as she alternately lowered her muzzle to reach aquatic vegetation and then raised it to monitor the breeze for signs of danger while chewing down a mouthful of weeds. She loved the buds and round pads of the yellow water lily and swallowed yards of their stalks, like strands of oversize spaghetti. Yet she rarely reached deep enough to get water in her sensitive ears. And when the breeze shifted, letting her catch a whiff of our scent, she faced us steadily, head up, ears forward, nostrils dilated, eyes searching for a recognizable outline. Although we made no move and stayed silent, she took no chances. With a snort that sounded more like clearing her throat than a signal, she summoned a small calf that we had not seen. It was hiding where she had left it, amidst a clump of bushes at the edge of the pond nearest us. Immediately it bounded into the water and swam to her, accompanying her out on the farther side. There she had a second youngster. The three vanished into the shrubbery and the maze of aspen saplings without even a backward glance.

A family of two is usual for a cow moose. Unlike the calves and fawns of other members of the deer family, young moose are born without white spots in their dark fur. Each calf stays for about two weeks in the thicket where its mother gave birth to it in late May or June. Thereafter it travels with her as a close companion until she is about to bear another family. The body of the calf takes on a well-fed roundness, making its long legs seem less awkward.

Through August and September the calf stays out of the way, close but uninvolved, as the tolerant bull that is silently accompanying its mother grows a new set of massive antlers and enters the period of annual rut. When his headgear is clean, he digs rutting pits and wallows in them, occasionally uttering a low-pitched whining sound that ends in an abrupt grunt. As her period of receptivity approaches, the cow moose wallows in the same pits while the bull looks on. He waits for her to signal by a moo-like call that she is almost ready for him. Then he begins to follow her closely everywhere, often leaning his neck across her back and stroking her with his dangling bell. If she walks too quickly, he grunts and she slows down. For several days he continues these attentions in a persistent pattern, interrupted at intervals when she stands still and he copulates with her.

Although moose are circumpolar, their family life shows distinctive features in the Old World and the New. In Eurasia, a big bull is likely to be herding several cows at once and to be the center of a harem-nursery group because the calves follow along. Less successful males would like to take his place, and he must defend it frequently. His territory is more extensive than that occupied by moose families in North America, where each bull tends to be monogamous and stay within an area only a mile or two across. Why these differences have evolved remains to be discovered.

Ordinarily we expect that monogamous habits and belligerent defense of the young by a powerful parent or two for almost a year will ensure the survival of a large proportion of the young. Yet wherever scientists have kept a close census, the mortality rate is surprisingly high among young moose between their birth and their first winter. An average of fewer than 50 from each 100 calves survives through autumn. The others have been abandoned when they failed to keep up with their mother. Some have become too entangled in the swamp vegetation and perhaps broken a leg. Others have drowned or have been fatally injured by lying down in their parents' path and getting stepped on. In Alaska, the calves often fail to escape the charge of a grizzly bear. Only under the best conditions do 25 out of each 100 calves live to sexual maturity.

Some female moose become pregnant when they are barely fifteen months old. In northwestern Ontario about 17 out of each 100 young cows bear calves of their own close to their second birthdays. Twin calves, however, are rare until a cow moose is at least three years old, when her growth is nearing completion. Young bulls, on the other hand, have little chance to participate in reproduction until their antlers have grown to generous size. In his second summer, a young bull produces a pair of spikes six to eight inches long. The antlers of the third summer have one fork, and those of the fourth summer a narrow palm with three or four upturned tines. Each new set is heavier and broader than the one of the year before, until at about age twelve years the bull is carrying around 80 pounds of antlers and frightening almost every animal in sight.

The wild moose whose private lives have been investigated most thoroughly are those that face only wild predators and other natural hazards in the forests of Isle Royale, a national park 48 miles out into Lake Superior from the northernmost tip of Michigan. The island is less than 10 miles wide and 45 miles long, but supports about 600 moose as counted in late winter just before the calves of the year are born. This stability of numbers, which matches well the food resources available to the moose, seems due to a fascinating combination of accidents, diseases, parasites such as worms, and attacks by about twenty wolves whose ancestors crossed the ice to Isle Royale from the Ontario shore during a particularly cold winter a few decades ago. The most successful wolves are the fifteen or sixteen in the main pack. Two lesser packs of two to three wolves each show signs of malnutrition and produce fewer pups. Together these wolves manage to kill and eat about 142 moose calves and 83 adults annually. None of the victims whose remains have been examined were between one and six years of age. The adults killed were mostly eight to fifteen year olds. Often their bones showed signs of debility.

For scientists the most intriguing feature of the moose and wolves on Isle Royale is the way the intermittent growth of the food plants is averaged out by growth in the moose population, which supports a steady rate of predatory and scavenging activity by the wolves. Despite the simplicity of the relationships, the situation seems to stay in balance, with almost no fluctuations from year to year. The herbivorous moose are the principal prey, the carnivorous wolves the only predators capable of bringing down a moose. On this sanctuary island, where no man interferes, an age-old drama is reenacted every year in a pattern that may have begun millions of years ago—before either moose or wolves emigrated across the Bering bridge to the New World and began to haunt the coniferous forests in America. Only the continued growth of the trees, reducing the amount of forage within reach of moose, appears to threaten the perpetuation of the system.

The Vociferous Porcupines of North America

Long before the moose and wolves reached the taiga country of North America, ancestral porcupines made their way to the same forests from northern South America. They ambled slowly, as all porcupines do, following the evergreen trees along the high crests of Central America, through Mexico, and into the southernmost Rocky Mountains. Now the porcupines *(Erethizon dorsatum)* are common into Alaska and Labrador, but remain absent from Newfoundland and the southeastern United States.

Every one of these quill-studded rodents of the forests seems to have an insatiable craving for bones and anything with a salty flavor. In Utah we have been awakened by a porcupine chewing on a fry pan that had been left at the campfire because it was still too hot to put away when we crawled into our sleeping bags. In Ontario and British Columbia, many an outdoorsman on a canoe trip has found his paddles and gunwales in sad shape from the strong teeth of nocturnal porcupines who liked the flavor of human sweat. In northern New Hampshire, porcupines sometimes attack tires and rubber parts of the hydraulic brake system of automobiles left out overnight because they are coated with salt from roads sprinkled for snow removal.

Otherwise the preferences of porcupines are far more like those of moose than might be expected in so different an animal. In winter a moose and a porcupine may be eating from the same evergreen tree, the one on needles within 12 feet of the ground and the other perhaps 40 feet higher up. The porcupine chews up needles and inner bark at its lofty position, not descending for several days if the weather is bad or if the snow is deep between one tree and the next.

In summer, the porcupine feeds more at ground level on herbs and low shrubs of many kinds. Sometimes it swims out into a pond, buoyed up by its hollow quills, and devours the fresh pads and flower buds of yellow water lilies. It might easily meet a moose with a taste for the same vegetation. It could also encounter a dog or other terrestrial animal that has not yet learned to leave porcupines alone. The porcupine's sensitive nose and shoe-button eyes recognize the danger. The quills rise as the muscle coat under its skin contracts, and the animal pivots on its front feet to present its back and tail to the intruder.

The quills on the back are as much as two and a half inches long, those on the sides and tail scarcely more than one inch, and those on the face and cheeks between three-eighths and three-quarters of an inch long. All of these quills have the same structure: hollow, multiply barbed close to the sharp tip, and loosely connected to the skin. Any animal touching the quills is almost sure to gather a number, as though they were burrs. Scratching the quills merely drives the points deeper. The barbs on each one prevent it from coming out again. The porcupine may also react to being touched by backing toward its attacker and swatting vigorously with its tail. One well-placed swat may drive 50 to 100 quills into the nose of a dog or deer that is overcurious. If given no first aid with a strong hand and a pair of pliers, the victim is likely to die of starvation, unwilling to eat while its lips are painfully studded with quills.

So pointed is a porcupine's defense that it is easy to forget the long, soft hairs that conceal the quills when the animal is not alarmed, and the fine, dark underfur that keeps its body warm in the coldest weather. The short-nosed rodent has no need to hibernate so long as it can find a den of its own or share one when winter storms arrive. The porcupine comes out after the sun is up, perhaps to travel a hundred yards along a trail it has used before. In late afternoon, as the light begins to fade, the animal usually heads homeward to any one of several hideaways, whichever is closest to a food supply.

When her young of the year is about to be born, a pregnant porcupine crawls into whatever shelter is handy. It may be no more than a cavity below a fallen tree trunk, or between the upturned roots of a blowdown. A thicket will suffice. Ordinarily she bears a single baby, to which Dr. Albert R. Shadle of the University of Buffalo has given the name "porcupet." Each one emerges in a protective envelope—the amniotic sac—but escapes promptly, with eyes open and most of its juvenile teeth already through the gums. It finds one of its mother's four nipples and nurses hungrily while its coat dries. Within two hours its quills harden and the little porcupet acquires its full ability to raise them in self-defense. Thereafter, when it wants a meal of milk, it must whimper to get its mother to sit back on her haunches and let her youngster reach her underside. There she has no quills and only short hair, which is poor protection if the porcupet should whirl and raise its armament. Perhaps this is why she weans it in a week or less and abandons it to its own fate in the forest.

A porcupet can climb two days after being born, but it moves with great caution and ordinarily does not ascend a tree more than half as far as its mother would until it is six months old. By then it is about 18 inches long and weighs close to four pounds. During each of the next two summers it will add another 4 pounds, and reach maturity when 29 months of age at a length of about 25 inches. This is still not full size, for some porcupines tip the scales at 40 pounds. These are solid-bodied, for the extra weight adds only about 5 inches to the length, not counting the 12-inch tail.

Each porcupine is as an individual. Some, as youngsters, are extremely playful, romping and tumbling about by themselves. Others appear stolid. Many produce little grunting sounds as they amble along, waddling because the front feet slant in.

Adults become extremely vociferous when September brings the porcupine year to mating season. They chatter like squirrels, meow like cats, bark like foxes, and intersperse between grunts in many keys a great variety of whines, moans, and grumbles. Females are especially noisy both night and day, and often seem to be carrying on a conversation with each male as he approaches. The male, however, keeps his distance—several feet at least. He has to have complete cooperation from her to avoid her quills while mating.

Her capitulation is sudden and brief. Quite abruptly she pulls most of her quills tight against her body, and lets others lie limp. Often she turns up her tail over her back, exposing its underside. The tail has no quills near its base, and does not endanger the male as he carefully comes close. Rearing up on his hind legs, he shuffles forward. At the last moment he may steady himself by resting his paws ever so gently on her back. He can give his full attention to the female and his own safety, for he has a baculum to stiffen his penis. Usually he copulates for only a few minutes at a time, as though aware that her mood will not last. Her state of readiness wanes in a few hours, not to be renewed for another several weeks. If she does not become pregnant on the first or the second period of heat, she repeats her performance—sometimes as late as January.

Western porcupines mate in September and October, eastern ones usually in November and December. The long gestation lasts 205 to 215 days. For no known reason, about 85 percent of the pregnancies are in the right-hand fork of the Y-shaped womb.

So modest a birth rate usually identifies an animal as having few enemies. This is especially true of porcupines that inhabit forests near human habitations, where many of the predators that can kill porcupines have been virtually eliminated. A cougar (mountain lion) can flip a porcupine on its back and bite into the unprotected undersurface. Somehow a great horned owl can manage this too. The most expert, however, is a smaller animal—the fisher—that trappers almost exterminated in the north woods for the luxurious pelts of the females. A fisher of either sex will climb a tree to meet a porcupine head-on. The quill-bearer can ward off this predator's attack only from a suitable refuge, such as the upper end of a deep gash in a tree trunk. There the porcupine can shield its vulnerable head and belly while keeping its hanging tail ready to slap back and forth at any attacker from below. A big old porcupine routinely used a wooden fortress of this kind in a New Hampshire woodland until continued damage to the tree killed it and hastened its downfall.

Foresters begin to complain that porcupines are too numerous and too devastating to commercial trees, when they see oval areas ("catfaces") cleared of bark on high limbs, or barkfree bands going all the way around the trunk, girdling it. Usually the woodsmen want prompt eradication of the porcupines, with no waiting for natural controls to take over. Occasionally the fact that fishers are already working on the problem becomes evident under the damaged tree in fisher droppings that have a few quills projecting, showing that they have gone through the fisher's digestive tract without causing the predator serious harm. Accidents and diseases due to parasitic worms or viruses seldom leave as convincing evidence that a population explosion among porcupines is correcting itself.

The Seals of the North Atlantic

Interactions between wild animals and humankind take many turns. In recent years, the number of gray seals *(Halichoerus grypus)* in the northeastern North Atlantic has increased at about the same rate at which people have abandoned the lonely islands. The change came slowly because the older residents chose to live out their lives, still content to accept a meager subsistence as a fair return for a large measure of independence. Their children went to school on the mainland and declined to return. More and more gray seals found privacy on the rocky coasts, as places to molt and mate and pup.

Grays are big seals, seven and a half to ten feet long as adults, often weighing more than 600 pounds. The increase in gray-seal numbers has had a marked effect upon fisheries, for the grays strongly prefer to eat salmon and sea trout. They pursue these important fishes and raid the nets that fishermen set. Moreover, gray seals spread a roundworm that parasitizes codfish, significantly reducing the value of cod the fishermen catch. The Scottish Marine Fisheries Agency calculates that nearly a fifth of the total annual haul of saleable fish from British waters is lost to the depredations of gray seals. This damage is concentrated around the Scottish coast because about 80 percent of the gray seals in British waters live there throughout the year. Only the penalities provided in the Gray Seal Protection Act, which was passed in 1914, and a continuing program to cull the gray-seal population keep the fishermen from waging war on the grays. Harbor seals concern them much less, because these smaller animals eat mostly fishes and shellfishes for which no commercial use has yet been found.

About 10,000 gray seals, constituting about a sixth of the world population, now come to North Rona, a little island between Skye and the Scottish coast. North Rona lost its last human resident in 1844. Almost as many grays find sanctuary in the Outer Hebrides, and still more on the bleak Orkneys.

On North Rona, the pupping and breeding season of the gray seals lasts from early September until mid-November. Repeated storms hamper attempts to keep a census. The grays avoid the sheltered coves because large numbers of harbor seals are already there, mating in the shallows and hauling out on land. The mature grays turn instead to exposed shores and creep well away from the water. Pregnant gray cows follow the biggest bulls in clambering toward whatever safety they can find. No one knows how many gray-seal mothers haul out no farther than the bare rocks nearest to shore to give birth, for the wild waves carry away the newborn pups to their doom. Any mother that loses her pup soon leaves.

Only a minority of the gray bulls survive the many hazards in turbulent seas into the years when they can serve as parents for their species. Many males get killed by violent water before finding a good route to the rocky shore. Once on land they fight viciously for a place to keep a harem. The winners, whose age falls between ten and twenty years, battle less among themselves than with younger aspirants in the age range from six to nine years. These challengers are sexually mature, but not yet strong enough to win and hold a territory.

A bull is regarded as successful if he has a harem of about 10 cows to supervise. Most males have 15 to 20 cows as proof of their extreme vigor, alertness, and pugnacity. Although the contests continue, the population ashore usually settles down to include about 26 younger mature males for each 10 harem masters, and one bull whose age lies between 21 and 30 years—a retired animal who is not yet senile. Females often live longer, to as much as 35 years, before they lose their ability to battle their harsh environment.

During the breeding season on North Rona, members of the Scottish Nature Conservancy make repeated visits to learn how the gray seals are doing. About one pup in six dies before the middle of October. Surviving pups to the total of 1,899 were still alive on October 24 of one year, each tended by a devoted mother who was already pregnant again. The following spring about 1,000 gray seals who were born the year before returned to spend their first birthday ashore. No doubt others had dispersed widely, for some with a numbered marker have been caught as much as 400 miles away from their birthplace.

Mature young gray seals, which are known as moulters, can be distinguished from the immature seals of both sexes by their lack of spots. The return of 1,400 in this age range gives no real measure of the natural mortality, for an unknown number remains at sea. The cows return more regularly after they attain sexual maturity at age five. More than 2,300 of them come to Rona to give birth and start off a new generation in the pattern that has become familiar.

Gray seals appear more leisurely than other kinds of seals in their swimming, their changes of pelage, and their family life. The pups nurse for about two weeks before being weaned, then fast for two weeks more while the white woolly coat with which they are born is replaced by a dark hairy one with almost no underfur. Meanwhile the mother has become receptive again to the harem master.

He recognizes the change from her behavior and her odor. Awkwardly lurching toward her because his flippers cannot support his great weight, he leans his neck over hers as though to whisper sweet nothings in her ear. This could be doubly difficult for him because her ear openings are inconspicuous and because his vision in air is so poor. Grays are "earless" seals (family Phocidae) with no ear flap—just a tiny opening that is often concealed among the wrinkles of the neck and can be closed tightly when the seal dives. Any seal on land is handicapped for vision because the corneal surface of the eyes is distorted, streamlined for swimming and seeing under water. It produces a poor image on the retina, ruined by nearsightedness and extreme astigmatism.

Ordinarily a gray bull treats his cows gently, despite weighing a third more than each of them. If possible, he nudges any cow that smells attractive into a shallow pool where the water will provide some buoyant support. Often the rocky shore offers no suitable pools and he must perform his mating on rough ledges. He does his best to prop himself up with one flipper and to hold her with the other. Generally he gets an extra grip with his teeth on a fold of skin at the nape of her neck. A female that is not ready can slip away, and some do as though they had changed their minds after signalling their willingness. The bull just shakes himself and shuts his eyes to snooze awhile. He is in no hurry. She is going nowhere until she is pregnant. His only concern is to prevent a younger bull from sneaking into his territory.

More than five-sixths of all the gray seals in the world follow this way of family life. They do not migrate. They do haul out once more each year for a few weeks to fast and molt. The cows change first, toward the end of January, and return to the sea by March, just as the bulls arrive for the same purpose.

The other sixth of the gray seals, totaling about 12,000 animals, is divided almost equally between a population in the brackish waters of the Baltic Sea and another in the North Atlantic from Ireland to Iceland to Greenland and down the coast to Nova Scotia. This minority, which cannot have occupied these regions for more than 9,000 years because of Ice Age glaciation, has changed its schedule of shore activities. They give birth and breed in February, and go through the molt and fast in September instead of winter.

The whole pattern is reversed, for the bulls precede the cows under the low sun that follows a short arc between dawn and dusk across the winter sky. The gray bulls search out suitable places to wriggle onto an ice shelf or an ice-clad beach, then progress caterpillarlike away from the water. Many roll onto one side and inch along, with no attempt to use their flippers. On good sites so many bulls congregate that no one of them can clear more than a few square yards around himself. Some appear to cease trying, and merely lean on one another. Fasting, the bulls wait despite gales and storms, in winter air that is bitterly cold—much colder than the adjacent water, since it gets some heat from the Gulf Stream.

Later, gray cows begin to drift along the same shores and pick out landing routes. A few cows approach each bull who has a territory. He makes room for them, even if to do so he has to roll and wriggle against his male neighbors, threatening them with bared teeth but rarely fighting. This is all the space the cows will have in which to bear and nurse their pups.

Late February must be a hungry period as well as a chilly one for most of the cows and all of the territorial bulls among the gray seals of the western North Atlantic and the Baltic. These adults remain on land, protecting the pups and keeping them supplied with a milk that is one-third butterfat. On it the pups grow rapidly while the adults lose weight. When the pups are two to three weeks old, their mothers abruptly wean them and come into heat. As soon as mating is successful, the adults depart for the sea. Behind them the helpless pups live on their fat while developing for several further weeks. Each pup must molt and decide for itself when to skid and bump down over the rocks and ice into the water. It learns to catch its own food there, or perishes in the attempt.

We naturally compare the family lives of the gray seals in their three areas with the behavior of the seal we know best—the harbor seal *(Phoca vitulina)* or hair seal, which Britishers know as the "common seal." We can see harbor seals through our binoculars almost any day in the year if we go a dozen miles to the Atlantic coast. Occasionally a seal of this kind swims into the estuary within two blocks of our home, and we notice the animal's round dark head as the seal swims among the small craft moored close to the town dock. Harbor seals turn up in such places along coasts of the North Atlantic from the Far North to North Carolina and to Portugal, and in the North Pacific from the Bering Strait to Korea and to Baja California. Some harbor seals even live in freshwater lakes, presumably where their ancestors became landlocked as the drainage system of the continent changed at the end of the Ice Age. Along Scottish coasts, the harbor seals bear pups and mate in protected bays just a week or two earlier than the gray seals busy themselves along the same rocky shores.

The harbor seal is smaller than the gray, and has a round face instead of a long one. It stays near shore all year, for it has difficulty sleeping in the water. At night it hauls out on remote rocks or mudbanks from which it can slither to safety. By day in fine weather, we see harbor seals basking in the same places and tolerating close companions of the same kind. While hunting for food, by contrast, a harbor seal prefers to go alone.

The season for pupping and breeding is longer for harbor seals than for the grays. According to geographic location and weather, the time comes and may continue for many weeks, between May and October. The customary place for giving birth is in shallow water. The buoyant pup floats vertically, with its head and shoulders exposed. It receives almost constant guidance from its mother for a few days, until it learns to swim well and gains strength. At first, the most the pup can do is sink into the water deep enough to nurse. For this it must hold its breath for minutes at a time.

Harbor seals go through peculiar antics for a few weeks before the pups are due and again for a week or so after the pups are weaned. The adults cavort in the water, seemingly as mated pairs. The earlier antics can be called "play mating"—after the two have been twisting and rolling together for a while, the cow turns over on her back while the bull climbs on her belly, clasping her with his fore flippers. After she has weaned her pup, however, the twisting and rolling game progresses to actual copulation. The adults molt at the end of the mating period, and spend as much time as they can ashore, fasting. They bark in protest if approached, and slither into the water to swim and dive until we go away.

The loudest calls are heard after the molt, when the bulls return to the water and begin fighting with one another. All through October and November their battles continue with no obvious reason. The usual centers of contention—the females—are pregnant and paying no attention.

Harbor seals are like the conventional grays in spending awhile ashore in cold weather, fasting and molting before spring arrives. Generally they wait until the March equinox approaches, and get it over with in record time. But one population of harbor seals has managed to change its schedule of birth and mating to April. These are the seals that live in the Gulf of St. Lawrence and along the coast of Labrador. They haul out on the ice shelves, often close to the more numerous harp seals. We hear little about them because, like all harbor seals, they lose their glamorous white woolly coats before birth, and wear only a thin covering of pale gray hair over their thick underfur.

Australia's September Oddities

Along many a road in eastern Australia, are signs warning "Koalas Cross Here at Night." Sometimes they cross by day also, giving a motorist a better chance to avoid the slow-moving marsupial or to intercept it and get acquainted. The bear-shaped animal *(Phascolarctos cinerens)* with its conspicuous hair-fringed ears and vestigial tail makes no attempt to bite if picked up. But the sharp claws on its five-fingered hands and four-toed feet can pierce the thickest clothing if the animal tries to grasp and prevent itself from falling. As we carried a koala to the nearest eucalyptus tree, we kept reminding ourselves that by staying off the ground in the forests, the animal avoids the two greatest hazards of its native land: drought and tiger snakes. We hope that the tree we picked for it belonged to one of the twelve species with leaves a koala will eat. In New South Wales the Blue Gum and the Gray are preferred, whereas in Victoria the Manna Gum is the favorite.

For most of the year the adult males and females slowly travel their separate ways. In early September, however, and until late January, the males show more interest in members of the opposite sex and follow them one at a time when their trails cross. Occasionally a male becomes so eager that he actually goes hunting for a mate. If the female he finds is not ready for him he waits, feeding close to her in the tree. When she is ready, he can dispense with any elaborate courtship, for she accepts him stolidly. Then he is off, perhaps to find another partner. Lucky males discover a whole treeful of females, which they guard zealously against the approach of any competitor.

The baby koala, no more than three-quarters of an inch long at birth, makes its brief appearance about 35 days after the mating, and travels under its own power into the pouch, which conveniently opens backward and contains two nipples. About six months later, the youngster is a "gum baby" about seven inches long and ready to get its head out of the pouch to look around. Soon the little koala begins clambering out and riding on its mother's back. It is not weaned, however, for a full year, and then by eating its mother's fecal wastes rather than eucalyptus leaves until its digestive tract can cope with the toxic materials that the fresh foliage so often contains. The young koala reaches full size at about four years of age.

The mother breeds in alternate years, probably from the time she is full grown until the end of her normal lifespan at about twenty years.

The family life of Australia's other oddity—the duckbill platypus *(Ornithorhynchus anatinus)*—remains far more difficult to observe, although the range of this strange egg-laying mammal extends from the vicinity of Adelaide east, north into Queensland and south into Tasmania. The duckbill spends most of its life in the scarce fresh waters of streams and lakes, emerging from its burrow in a bank in the early morning and late afternoon to grub for food in the bottom sediments of shallow places. At the most, the body of a duckbill is eighteen inches long, and its broad furry tail another six inches. Its tail, the webbing on all four feet, the flat naked beak (which resembles that of a duck), and the groove on each side of the head that closes to shield nostrils and eyes when the animal dives, distinguish the creature from any muskrat.

Adults of both sexes share the same burrow except during the breeding season. Then the female leads the male on a merry chase, like an underwater ballet, as he tries to swim fast enough to catch her. He wins after he seizes the tail of the female in his forepaws and lets her tow him around awhile in circles. After mating, the male returns alone to the home burrow while his mate digs a new one as a nursery. She lines it carefully with wet

Duckbill platypus *(Ornithorhynchus anatinus)*

leaves, which she carries between her body and her tail by bending the appendage forward below her. Her webbed forefeet are used to squeegee the leaves and her fur, keeping the nursery as dry as possible.

About two weeks after breeding, the female enters the nursery and plugs the entranceway. She lays from one to three eggs, each soft-shelled, almost spherical, and half an inch in diameter. For about ten days she either curls her fat body around them as well as she can, or lies on her back and incubates them atop her warm belly. She lacks a pouch in which to give them better protection and nipples from which the young might nurse after they hatch.

At first a baby platypus measures no more than an inch in length. It is naked and blind. It licks milk from fur over the glands on the abdomen of its mother, and grows on this diet to be fully furred and thirteen inches long within four months. The youngster can then follow its mother into the water and encounter its share of Australia.

Young duckbills must avoid any large snake or lizard, such as the carpet snake *(Python)* and the goanna *(Varanus)*, which hunt for prey at the water's edge. Today this is easier than in the past, for most of these predators have been destroyed by the artificially introduced European rabbit. Duckbills do get caught inadvertently in snares set for the rabbits or in traps set for freshwater fish. Together these hazards keep the extraordinary mammals scarce, despite Australian laws that prohibit exploiting them for their soft brown pelts.

CHAPTER EIGHT

The October Set

ORBITING SPACECRAFT KEEP constant watch on the earthly home of mammals and humankind, photographing the cloud cover and reporting on surface temperatures. Technology keeps current our awareness of the world and of the interrelated events on its surface. Climate has no geographical boundaries. Knowing how much October in the Southern Hemisphere is like April in the Northern, we feel encouraged to take a global view.

Observers on the ground fill in the details of seasonal reactions in local animals. The caribou of arctic America move south, impelled by the first snowstorms of approaching winter and by lengthening nights. Each October day may be twenty minutes shorter than the one before. The caribou are in fine shape from a summer's feeding, and the bulls find the cows more receptive to amorous advances than in any other month.

On the arid deserts of the American Southwest, the pronghorns are similarly occupied. They carry on a tradition they formerly practiced across the broad center of the continent, from Canada to Mexico and from eastern California to the Mississippi. Only two centuries ago, the pronghorns numbered between 20 and 40 million animals. By 1908 less than 17,000 were left. Now there may be 350,000.

In wetter parts of the Northern Hemisphere, we bless the autumn rains. They refill the soil with moisture, soak the fallen leaves, consolidate them, and speed their decay. The water sustains the overwintering roots and bulbs, and ensures that deep wells will yield plentifully through the cold months to come. It flushes out the stagnant pools, and clears the rivers of debris.

In southern Europe and Asia Minor, the autumn rains drive mole-rats to slightly higher ground, where they begin building breeding mounds. Far south in Africa, from the Cape of Good Hope to Zimbabwe-Rhodesia, unrelated mole-rats respond similarly to the wet weather of October. Both types of animals start families in time to take advantage of plant growth that has yet to show.

In Australia, the gray foresters, which are the giants among kangaroos, rut more regularly in this month than at any other time of year. This is not merely an activity of the South Temperate Zone, for the foresters follow the same schedule far north in tropical parts of the Australian state of Queensland. Similarly, in Patagonia the big-eared rodents known as viscachas (or Peruvian "hares" despite their long, curled tails) leap over their rugged, rocky landscape—male after female—in springtime abandon. Along the Andean slopes, the mountain viscachas synchronize this activity as close to the equator as Bolivia. These are ways of life that have been inherited, and they are observed strictly despite differing cues from the environment. Even the advent of humankind and the conversion of so many areas to human enterprises have not upset the calendar these animals live by.

The Caribou and Reindeer

The most northern deer are also the most democratically decorated. They form one circumpolar species *(Rangifer tarandus),* as caribou and reindeer. Largest of them are the woodland caribou, weighing 700 pounds in October. They stay in the northern coniferous forests of America. The slightly smaller barren-grounds caribou shelter for the winter in the same spruce-fir forests across North America and in Siberia, but migrate northward to seashores of the Arctic Ocean when spring arrives. Smallest are the reindeer of northern Europe and western Greenland, which herders keep moving lest the animals destroy the meager vegetation.

Reindeer and caribou are unique in the deer family in routinely growing antlers regardless of sex. The rack on a cow's head may be almost as magnificent as that on a big bull. She retains her antlers longer than he does, until about a week after her calf of the year is born. The calf too will possess antlers before it is six months old.

Except in October, the woodland caribou tend to be elusive and solitary. A bull or a cow with her calf each need a square mile or two of coniferous forest and bogland in which to roam and feed. This large expanse is necessary because so great a proportion of the plant material present is unacceptable as nourishment. The conifers and shrubby members of the heath family have scant appeal. Nor will a caribou eat many ferns or the peat moss that chokes the areas of bog water. The attractions instead are the seedling birches, the scattered grasses, and the lichens that carpet the drier ground between the trees. Winter snow often makes these foods hard to find.

Fortunately, the caribou calves wean themselves early and let their mothers begin to put on weight during the summer months. Each female will need her reserve of fat during the winter ahead. The bull caribou has an even longer season to feed exclusively for his own good. By October he is in prime condition, with a magnificent winter coat of brown hair set off by a ruff of white around his chest and neck. To a varying degree the white extends back over his shoulders, almost to his hips. His feet seem adorned with white spats. His white tail shines brilliantly. His antlers spread backward and outward, each as a somewhat flattened beam bearing even more flattened branching tines. One of these is the peculiar brow tine or "shovel" that extends forward just above his nose.

Some of the biggest bulls cannot wait for October to begin rounding up a harem. Before September ends they may have driven half a dozen females into a manageable group and still be eager to triple the number of their cows. Each bull is so busy rushing around his harem, threatening any other bull that comes near, that he has no time to eat. He gets his energy from a pad of fat over his hips. It may be three inches thick on October first, and almost completely gone a month later. By then, however, he will have had his chance with each cow in turn as she came into heat.

A bull makes no attempt to keep his harem in any particular area that might be regarded as a territory. Instead, he provides a shield around a group of females and tries to limit their tendency to disperse as they search for food. As Yngve Espmark notes from his studies of reindeer near his base at the University of Stockholm, each bull makes his mobile harem the center of his territory and himself the most conspicuous individual present. When his sense of smell tells him that a particular cow is coming into heat, he stays close to her as much as possible. He attacks bushes in her immediate vicinity, slashing at them with his nose in sweeping arcs that display his rack of antlers. The brow tine deflects many of the branches that otherwise would strike his eyes, for he seems oblivious to the damage he might do to himself. At the same time, a secretion from glands on his feet becomes smeared on the shrubbery, adding an odorous message to the commotion and to the grunting sounds with which the bull demands attention. He may approach the cow repeatedly and sniff at her from as close a range as she will allow. But until she signifies her readiness by nuzzling at his flanks he makes no attempt to mount her. Whether from

Caribou (*Rangifer tarandus*)

impatience or suppressed excitement, he urinates at frequent intervals during the enforced wait. Some scientists believe that the odor of his urine, which is obvious to a human nose, has an aphrodisiac effect upon the cow. Certainly he splatters the liquid freely over his own legs and all of the adjacent shrubbery. The cow too urinates copiously as soon as the bull has finished copulating with her. Abruptly she loses her attraction for any bull.

The caribou on the open tundra remain far more sociable throughout the year than the woodland race. In compensation, they must keep moving to reach areas of their range where the plants have had a chance to grow since the last visit of a herd. The birches, grasses, and sedges, which together constitute almost 70 percent of the summer diet, recover more quickly than the shrubby lichens that account for 30 percent in summer and more in other seasons. The lichens known as "reindeer moss" are particularly vulnerable because of brittleness in dry weather. At least ten years should pass before caribou return to feed in any area.

The reindeer herders in Lapland and adjacent parts of Scandinavia are an artificial substitute for the natural nomadism that ancestral reindeer must have shown prior to domestication. The best that the tundra can produce during its short growing season is essential for the growth of calves and the recovery of adults that have survived the long, bleak winter. More bulls than cows die of malnutrition before spring arrives, seemingly because they have fasted and depleted their reserves of fat during the previous rut. Perhaps this is nature's way of controlling the population by eliminating the less resilient bulls, conserving the limited supply of food for other mouths.

The reindeer and the barren-grounds caribou that do survive to see another springtime soon regain their strength. They put on weight despite internal parasites and the unrelenting attacks of blood-sucking mosquitoes and blackflies. Each pregnant cow interrupts her browsing for about 30 minutes to give birth, to clean her calf thoroughly with her tongue and learn its distinctive odor, to eat the afterbirth, and to wash herself. Then she nudges her precious calf to its feet and, with it at her heels, moves on in the browsing herd.

During the southbound trip in early autumn, the caribou and reindeer travel in closer formation than on the northbound routes, when they are hungrier. Their tighter proximity increases the likelihood of careless collisions and antler-to-antler contact. Apparently this has an erotic effect at any season. For bulls that are well fortified with fat, it stimulates the secretion of sex hormones. The older, bigger males begin their attempts to form a harem. By October, the cycle of the year has run its course. The herds have reached the edge of the coniferous forest and the shrubs that the bulls can fight in ritualized display.

The success of this schedule should be measured in nature's way: by the number of extra lives that can be sacrificed while keeping the population constant. These extras hold importance in the nourishment of wolves and Eskimos or Lapp herders. For the surviving caribou, which are the ancestors of all there will ever be, to be well fed and sexy in October may be more important than the shape of an antler.

The Swiftest Mammals in the New World

The first pronghorn buck we met raced alongside our moving automobile on a gravel road in Arizona, and then sped past the car while the speedometer held at 35 miles per hour. He accelerated a little more, crossed deliberately in front of us, and stopped, letting us go by. Through the rearview mirror we saw him casually grazing, his contest won, and his interest in us completely ended. Many a traveler on back roads in the West has enjoyed a similar experience.

Top speed for these exclusively American mammals (*Antilocapra americana*) is probably not much more than 40 miles per hour, although estimates of a mile per minute are often bandied about. To us the most charming feature in the natural prowess of the pronghorn is its apparent enjoyment of competitive racing, whether with another pronghorn, a man on horseback, or some vehicle of man's devising.

Only the bucks that have not yet had a share in starting a family seem to engage in racing. They band together in loose bachelor herds for much of the year, and become interested in does chiefly when October arrives. The older bucks are fatter and more settled. They stay with the does throughout the year, and have vested interests when the does come into heat. Usually a buck in his prime has no need to battle seriously to keep a harem of two to four does. A bigger harem presents problems, for the females tend to wander off in every direction—and to be driven away by wandering bucks. Probably any doe can outdistance a fat buck if she wants to. Rarely is one sufficiently coy or temperamental to try. But generally each doe tests the ardor and persistence of the male by keeping just a few steps ahead for hour after hour before she lets him have his way.

Often we hear people refer to pronghorns as "American antelope," as though they were a counterpart to the fleet antelopes of the Old World. Actually the pronghorn is unique, the sole survivor of a separate family that zoologists regard as intermediate between the deer (family Cervidae) and the bovids (family Bovidae) such as antelopes and cattle. Like no other animal on earth, a pronghorn of either sex produces a pair of head ornaments with a bony core that keeps growing and a sheath that is cast off annually. The bony core is like a true horn, but the sheath is composed of fused hairs and is quite unlike a solid antler.

Shortly after the rutting season ends, the old black sheaths over a pronghorn's armament loosen at the base. New stiff hairs begin to grow up underneath. Gradually they push off the old sheath and take its place, first as a thick, soft membrane and then as a horny layer that hardens progressively from the tip toward the head. On a doe, this change is not particularly obvious, for her horns are rarely longer than 5 inches and show only an indication of flattening and forking toward the tip. On a buck, the horns grow more conspicuous every year. At the end of his second summer they may be 2 to 3 inches in length, and 12 to 13 inches by the time he is five years old. In 1899 a buck was caught with horns measuring 20 5/16 inches along the outer curve; the outer tips of this record pair were 16 3/16 inches apart.

Today the number of pronghorns in America is about a hundredth of the population two centuries ago. The biggest herds are still west of the Rocky Mountains, in California, Nevada, and eastern Oregon. They are easier to see, however, in Wind Cave National Park, South Dakota, in the Custer State Park nearby, and on the National Bison Range close to Moiese, Montana. In these areas their food habits and family life are becoming better known.

The pronghorns are like many (perhaps most) other animals in facing their greatest challenge to survival when winter comes. Cold alone is not the hazard, for their handsome fur coats keep their bodies warm even in severe weather as long as they are well nourished. But from November on, they require plenty of sagebrush and creeping juniper, both of which retain their leaves all winter. If the supply of these evergreens is limited and the pronghorns must subsist on dead grasses, they are in trouble. The fawns of the preceding summer succumb. So do large numbers of the mature bucks—animals four and a half years old or more. The pregnant does lose or absorb so many of their unborn young that the birth rate falls drastically the following spring. As few as 40 young from each 100 does may be born, instead of the customary 90 to 220.

Those that survive until spring become less sociable because of hunger. They spread out over the territory available to them and graze on the green grass spears that come up so abundantly in April. In May they transfer their attention to the tender, broad-leaved herbs, and in August to the foliage and shoots of deciduous shrubs, which satisfy them through the October rutting season.

As September ends, these handsome animals begin to congregate in little groups, partly through the efforts of the mature bucks. By then these males have reached some agreement on territorial boundaries, and each buck claims an area of 40 to 80 acres. His borders overlap with no neighbor's, but between the territories of one buck and the next there may be a no-pronghorn's land in which neither buck will tolerate intrusion of a yearling male. Whatever does and fawns stay within the boundaries are herded together as much as possible, and become the harem the buck will serve. He does not overlook a fawn that comes into heat at less than six months of age, and many a yearling doe gives birth as proof of his attentions.

A newborn animal of this kind seems to wilt against the ground. It holds its ears flat against its head, and shuts its big eyes if any creature other than its mother moves nearby. Ordinarily a doe stays within 100 yards of her offspring, but leaves twins (if she has them) separated by 30 feet or more. When no danger is in sight, she cautiously approaches one and then the other to let them nurse for a minute or two at a time. Then down they must go into their inconspicuous crouch until, at about one week of age, they can race after her at 25 miles an hour if she heeds a warning.

The warning signals of pronghorns are mostly silent. Each animal makes full use of muscles in the skin to change the angle of hairs in the coat. Those on the rump turn aside the tan hairs and abruptly expose a flare of glistening white. Any pronghorn seeing this sudden display turns toward it and repeats the signal like a relay center on a telegraph line. In mere seconds the visible warning travels a mile or more, alerting almost every pronghorn in sight. The system cannot be perfect, for some pronghorns will have their heads down, feeding, and not notice the flashing rump patches on distant neighbors. A backup system provides for this eventuality: in each white patch is a gland that releases an odorous material as soon as the muscles erect the hairs. Human noses can recognize the scent at a quarter of a mile, and pronghorns surely must be even more sensitive to it. Any animal that has its head down reacts at the first whiff. It stands up straight, stares alertly to see if any white rump patches show, and repeats the signal to other pronghorns downwind.

So far the wealth of chemical communication among pronghorns remains largely unexplored, yet the basis for a remarkable repertoire is known. Each individual has odorous glands not only on its rump but also on the lower jaw, at the base of each horn, on the lower back, and near each hind foot. Additional glands between the two parts of each hoof leave a chemical trail for any passing pronghorn to read. Unfortunately, science still lacks a means for observers to magnify and recognize a scent at a distance. No one has learned, for example, whether a mature buck who turns broadside toward a rival on the next territory is also sending an olfactory dare. When he fights a sham battle with a shrub, he probably anoints it with secretions that any pronghorn can interpret. We simply do not know as much as we would like about the swiftest mammals of the New World.

The Mole-Rat Way of Life

A rat-sized rodent that tunnels underground like a mole is almost automatically called a mole-rat, but this vernacular name has been applied to unlike animals of four different families native to various parts of the Old World. We have yet to meet the mouselike kinds (family Muridae) called bandicoots *(Bandicota)* in Ceylon, or the hamsterlike ones (family Cricetidae) known as zokors *(Myospalax)* in northern Asia. More familiar to us are those of Israel and adjacent parts of the Near East (family Spalacidae; genus *Spalax*), and the "common" mole-rats of Africa south of the Sahara (family Bathyergidae, particularly *Cryptomys*), both of which build breeding mounds in the middle of October when the rains have forced them to congregate on higher ground.

Our first encounter with a mole-rat came during an all-day drizzle near Beersheba, close to the old boundary between Israel and the Negev desert. We knew that the rainy season had begun in October and would last another month. By February, five months of mild winter would bring virtually all of the annual rainfall, which averages ten inches at Beersheba. We could scarcely believe that subterranean animals in such an arid land are in danger of being flooded out so regularly that they routinely migrate to higher areas right after autumn equinox. Yet Ziv Shiftan, the Israeli geologist who accompanied us, insisted that this was so and drew our attention to some small hills of earth half hidden among the grasses and low, stiff shrubs. Some hills rounded up fully three feet above surrounding contours, but seemed to be centers from which lesser mounds radiated in broken lines for fifteen feet or more.

We dug into one of the central hills with a shovel and found a labyrinth of interconnected galleries. Some, almost three inches in diameter, showed by their hard-packed surfaces that they were much-used hallways.

They interconnected larger chambers, most of which contained a thin mat of plant fibers but otherwise proved empty. One room appeared to be a larder, in which a dozen succulent bulbs were stored. Another clearly served as a privy. Each burrow system is the private world of one mole-rat, which digs and maintains it, using its teeth to loosen the earth and its snout to shift the debris and pack the walls, floor, and roof until they are unlikely to erode from traffic.

We finally caught one of the animals under a bulging ridge where the tunnel roof was less than an inch thick. The cylindrical body showed no indication of either tail or eyes, and was covered by short fur suggesting that of a mole because it could be brushed easily in any direction. Gray-black in color and about seven inches long, the muscular animal twisted vigorously and tried to bite in self-defense. Its lips pulled back over two large incisor teeth above and two below that protruded from its powerful jaws and showed that it was a rodent. The broadly rounded snout, with deeply cleft nostrils, seemed firmly padded and covered by tough, leathery skin. The short legs ended in surprisingly delicate feet with small claws.

A drizzle dampened our enthusiasm for further digging, which was just as well since we might easily have cut into a nursery chamber. Young Palestinian mole-rats *(Spalax ehrenbergi)* are born in February, or in various weeks from December until March, in a single litter of two to four. At first they are naked, pink, and helpless, and about two inches long. Long gray fur grows out within the first two weeks, making the youngsters appear quite unlike the parent although they similarly lack a tail and eyes that can be seen through the skin. Juvenile mole-rats leave the nest when about four inches long and five weeks of age. Sometimes they wander quite extensively each night, while choosing a new homesite.

During the dry summer, the mole-rats dig rather simple tunnel systems to reach bulbs, roots, tubers, and other subterranean parts of plants. Occasionally the animals come out for some nocturnal foraging on foliage, fruits, and seeds. Perhaps they learn something about the topography too, for when the October rains begin they move to the nearest high points and construct their personal shelters. Each mole-rat seems to need several days to settle down in its new quarters, and seems slightly less eager for a fight if it meets another member of its own kind. Winter is the only season when either sex shows any tolerance for the other.

When a male and female meet, they first stand nose to nose, growling and drawing back their lips to expose their formidable incisor teeth. Even if a female changes her voice to make high-pitched cries like those of young mole-rats, the male may still attack and bite. She may stand her ground, or run away while the male pursues until she whirls and again confronts him nose to nose. In encounters that eventually lead to copulation, the attack phase lasts an average of eight minutes or more, then changes dramatically to courtship.

A courting male softens his calls until they are barely audible—almost a trembling whimper. He lowers his padded upper lip, thereby concealing his dangerous incisors, and begins to nuzzle the female. Usually she responds at once, showing no resentment for his previous rough treatment. She lets him caress and lick at her head and body, push against her, bite gently and pull at her skin, or even step over her head. She may lick at the male, while continuing almost without interruption her high-pitched courtship cry. Depending on the individuals, these preliminaries continue for 8 to 60 minutes before both partners are ready for copulation.

The first few times that the male climbs atop the female's hips are practice runs. But the excited female keeps urging him to return, calling her courtship notes over and over. Again and again he resumes his mounted position, each time making a little more progress. Suddenly the lovemaking ends. She falls silent while he moves away a short distance to clean himself. When next she uses her voice, it will be an utterly different call: that of the pregnant female. It is bell-like, low-pitched, and tremulous.

For a month the male may stay with his mate and help her enlarge the breeding mounds and radiating runways. But he leaves—or she ejects him —before her young are born. Thereafter he is a stranger to her, and she to him. Each seems as insistent on solitude as before October came.

Dr. Eviatar Nevo, a professor of genetics at Hebrew University in Jerusalem, became interested in mole-rats more than a decade ago when he discovered a correlation between the complexity of the breeding mounds and the height of the water table in the soil. To his surprise he found four different true-breeding types of the Palestinian mole-rat, each with a distinctive number of chromosomes (52, 54, 58, or 60) and a separate geographical range. Courtship patterns varied in intensity and duration, and hybridization seemed unlikely. The type with 52 chromosomes lived in the mountains of Upper Galilee. The one with 54 was characteristic of the Golan Heights and Mount Hermon. Those with 58 tunneled in central Israel. The remaining type ranged over the northern Negev. Probably each was a species that had been in the making since the Ice Age. Their isolation is difficult to understand in a country that is smaller than the state of Massachusetts and in which the mole-rats find suitable conditions only in that third where rainfall is greater than four inches annually.

The Sunbathers of the Andes

For fully 2,800 miles north and south along the mountains of South America, the hare-sized rodents known as mountain viscachas (*Lagidium* spp.) find den sites among large tumbles of boulders. These animals, whose name the Spanish explorers transliterated from an Indian word that sounded like a sneeze, do not dig burrows. Yet they must have protection, especially from the long-legged Andean foxes *(Dusicyon)*, and close enough access to water to avoid hazardous trips across open country. They find these amenities and acceptable plant food at elevations as great as 17,000 feet near the equator and at lower levels where the mountains dwindle away at latitude 52° in Patagonia.

Soon after sunrise, the viscachas climb out of their dens to perch on favorite rocks and ledges. With tails uniquely curled upward, they dress their fur and stay on the alert, their large ears raised high and eyes wide open, watching for signs of danger. If nothing alarms a viscacha, it settles itself comfortably for a sunbath that may last all day. Yet in any colony of these gregarious rodents, some individuals always have their eyes at least half open. Every viscacha responds fully to any suspicious sound. Neither a fox nor a hawk will be overlooked for long. A man, particularly if he is accompanied by a dog, causes an immediate uproar. One viscacha whistles and every head comes up. The alarm call continues for about a second. It may be repeated three or four times at intervals of less than a second and a half. Then every viscacha dives for its den, although it may wait on its doorstep for a minute or so before vanishing into the shadows below the rocks.

Late afternoon is meal time. As the sun gets low in the sky, one viscacha after another stretches, then hops slowly like a rabbit or a small kangaroo to some favored vegetation. Its tastes seem catholic, for it will eat its fill of the leaves and bark of the tola shrub *(Lepidophyllum)*, the bark and blossoms of groundsels *(Senecio)*, or the blades of fescue grass *(Festuca)*, nibbling each leaf lengthwise as though in no hurry. As soon as darkness gathers, the viscachas disappear into their dens. Often they go no great distance for, if a flashlamp is used to illuminate their retreats, the viscachas' eyes glow bright orange-yellow. No sign of bedding litters the floor: just a few broken whiskers, and perhaps a pellet or two of dung.

Paradoxically, the breeding season in October is the time of year when mated pairs are least often seen together. During late September, as her period of heat approaches, the female becomes extremely belligerent and refuses to have her mate or any other male in her vicinity. Some of the males move out altogether and establish themselves in temporary bachelors' quarters. Other males stay within sight. By being persistent they notice a gradual change in the females' reactions. At first a female will allow a male to approach her from in front, perhaps to nuzzle her head. If he tries to go around her, she rushes at him, squeaking loudly until he departs. Then, for less than a day, she responds quietly and gently to her mate. Now it is the males' turn to be temperamental. Some become extremely possessive, squabbling with any other male that comes close and copulating with the one female at frequent intervals. Others prefer to be promiscuous, and mate with every female that will accept them. In some instances, a male with monogamous behavior is so permissive that visiting males seem welcome to copulate with his mate. However, sexual activity soon ends for each female because components of the seminal fluid congeal to form a soft, white waxy plug about two-thirds of an inch in diameter and more than an inch long. It effectively fills her vagina and does not drop out until some time the next day. By then her sexual receptivity has waned and she will squeal if mounted, rather than whimper in acceptance.

Curiously, only the right side of her Y-shaped womb ordinarily functions, and after mating a single embryo begins its slow development. After five weeks, it is no more than an eighth of an inch long. Her baby is not born until January, when it weighs less than half a pound. It is fully furred, however, and surprisingly independent of its mother. Its teeth are functional, and it can eat plant food. Her milk is just an easy supplement. To nurse, the youngster sits upright alongside its mother while she is in the same position. It can then easily reach one of her two nipples, which are far to the sides of her chest. Occasionally a mother nurses two youngsters of different ages at the same time, but whether one is adopted or an offspring from a previous mating is not known.

As many as 75 viscachas may live in the same rocky area. Each member of the colony, each pair, or each family has its own den and sunning place. Trespassing is rare, and so is outright fighting. These sociable animals have overcome most hazards of their own making. Their chief dangers, other than predators and infections of various kinds, are in a sudden change of weather—a rain squall, a hail storm, or a snowfall—that would wet them to the skin before they could scramble to shelter. Viscacha fur has few guard hairs to repel moisture, and gives no protection from the almost constant cold if wetted. Safety consists of staying home until the snow or hail melt and the moisture runs down the mountainside. If this takes several days the viscachas fast, for they store no food to tide them over during bad weather.

The numbers of viscachas have probably changed little from former times because the principal limitation affecting them is the availability of rocky areas in the mountains. Their pelts attract no furrier's attention, if only because they seem to be forever shedding in patches. No domestic animal is threatened by viscachas. Even the shrubby plants, grasses, and lichens they eat are so close to boulder-strewn slopes that no person has any use for the vegetation. The viscachas are protected from being taken as game by an old Andean superstition that eating the flesh of viscachas will turn one's hair white.

Boomers, Fliers, and Joeys

Australians say kangaroos have a five-footed gait. When traveling slowly, these amazing animals touch down with their small front paws while the sturdy tail rests on the ground and the two powerful hind feet move forward. At higher speeds they become bipeds, and hop as much as 30 feet at a bound while using the tail to counterbalance the front part of the body.

The biggest of the kangaroos are the gray foresters *(Macropus giganteus)*, of which a full-grown male, called a buck or "boomer," may measure 60 inches from the tip of his nose to the base of his tail and another 43 inches to its tip. He will weigh close to 150 pounds. Slightly smaller when full-grown, the females, called does or "fliers," move about with their young, or "joeys," in "mobs." Some joeys are still in the pouch, while others hop along after their mothers.

By choice the gray kangaroos remain in the thin shade of eucalyptus forests. Above them rise tier after tier of slender branches bearing narrow leaves that hang edgewise to the sun. Light streams through to seedling trees and many lesser kinds of vegetation, including the grasses that attract the kangaroos. They prefer to feed at night and rest by day, although no native animal in Australia today will pounce on a mature kangaroo at any hour.

Despite the remarkable tolerance of individual eucalyptus trees for drought, forests of these trees depend upon a climate that brings some rain almost every month, and a year total exceeding 30 inches or 40 if the air is hot and dry. This requirement is met chiefly in the southwest corner of the country, in Tasmania, and near the Pacific coast. Forester kangaroos abounded in all of these areas until colonists moved in. By 1888, a century after the settling of Australia began in earnest, the grays had been dispossessed in more than a third of the potential forest land in Western Australia and New South Wales. Kangaroos were killed off at every opportunity for sport and to protect crops on the new farms and grasslands that had been substituted for the native vegetation. Now wild grays are hard to find. In Queensland, where about 500,000 were killed each year from 1960 to 1965, the men who hunt them at night with dazzling lights, repeating rifles, and refrigerator trucks for the sake of the fur and pet-food industries complain that reproduction is not replenishing the supply.

The gray kangaroos' mating activity reaches a peak in October, although some continue producing young all year. The females are the bottlenecks, for they are equipped to add only one joey annually, or to mate again in thirteen days if something happens to their baby before it leaves the pouch. A boomer is ready to court a flier almost any night or day from his third year on, until old age closes in on him at about fifteen years. Yet he remains a transient, with no harem or territory. He joins small mobs, sometimes battling with a male that is already there. His immediate goal is to come as close as possible to each female, to sniff at her pouch, urine, and body openings, and to clutch at her tail while he learns her sexual condition. Neither her size nor age concern him in mate selection, for a young female becomes sexually receptive between her seventeenth and twenty-eighth month. Once mature, she will be ready to receive a male every thirty-fifth day.

Probably most female gray kangaroos develop slowly, for those two years old are seldom pregnant. The rate of pregnancy is close to one in twenty. However, the probabilities reverse between four and fifteen years of age, for in this range about nineteen in twenty are pregnant. Fliers that live longer than fifteen years have less chance to raise another joey. The pregnancy rate falls to 60 percent by age sixteen, to 40 percent by eighteen, and to zero among the few kangaroos that survive to age twenty.

The courting boomer is more persistent than romantic. He trails along after any flier whose odor attracts him, or approaches her hind quarters. Sometimes he touches her gently with an outstretched paw. He will groom her fur if she lets him, and lick her face if she turns around. His soft clucking calls are for her ears alone. Even when her odor tells him that her time has come, he must wait until she discovers this fact for herself. Her sturdy tail blocks his approach for copulation until she raises it and lowers her body by bending her knees. Only then can he enter from behind and fertilize an egg.

A pregnant flier shows no bulge that tells her condition. The embryo inside her womb is less than an inch long and adds just a twentieth of an ounce to her weight. Yet from it come chemical messages that alert the mother about 33 days after her successful mating. Without guidance she hops over to a big tree, sits down with her tail between her outstretched legs, and leans back against the trunk. She bends nearly double to thoroughly cleanse her pouch with her tongue. She polishes its smooth lining until it is pink and spotless. She washes the four slender nipples that hang down from the bare skin that ordinarily is hidden within the pouch. Then she leans back and waits.

The first sign that anything more is about to happen is several drops of fluid dripping from her birth canal, followed by a clear spherical sac almost an inch in diameter. It contains wastes from her unborn baby, and falls untended to the ground. Fastidiously the mother cleans away the residue of fluid and waits again. Soon the baby emerges, head first. Until it is free of the pressure of her birth canal, its amniotic sac encloses the infant. After it wriggles clear it begins a slow climb through the tangle of fur between her birth opening and her pouch.

Curiously, the mother seems to ignore her struggling youngster. She concentrates, instead, on licking away the additional drops of fluid that emerge from her birth canal. She captures the small mass of afterbirth with her lips and swallows it. Her tongue keeps busy in the fur her baby has already passed through, but never touches the youngster's body or clears any path to ease its progress. By its own efforts it must reach the

Gray kangaroo (*Macropus giganteus*)

pouch, climb inside, and find a nipple. Stretching its mouth as wide as possible, the little kangaroo works itself onto the nipple, swallowing forcefully until the tip is in its stomach. Now it relaxes, probably exhausted from so much effort, and just sucks away at the clear, fat-free milk its mother supplies copiously.

Until a newborn kangaroo has had a few hours to suck on its mother's nipple, it can be slid off if handled gently, and slid onto another nipple. It can even be transferred without harm to the nipple of an unmated flier if she was in heat 33 days earlier; her virgin nipples offer the same clear milk, and she reacts to having a baby on one of them just as if she had given birth. First the tip of the nipple enlarges, forming a little button in the stomach of the baby, keeping it securely attached to the mother until it grows big enough to let the button slip through its gullet. No wonder the Dutch explorer and merchant Francisco Pelsaert claimed in 1629, on the basis of his examination of the young in a kangaroo's pouch, that the baby grew from the teat in a unique kind of sexless reproduction!

Most joeys remain inside the pouch until they are about 190 days old. By then they have made some limited acquaintance with the outside world by sticking out their heads and sometimes their paws too. They reach for and nibble on bits of plant material upon which their mothers are feeding. Then the youngsters begin hopping out and exploring a while before diving back in head first. Even after they leave the pouch permanently, which may be as late as 315 days of age, the mother still welcomes her joey back to get milk. The young-at-foot is never weaned abruptly. It returns to the elongated nipple to which it was originally attached. This one, unlike the other three, now produces a thick and creamy milk. The consistency has changed greatly during the pouch life of the joey.

Mother gray kangaroos differ greatly as individuals in their reaction to a boomer. Some will mate while pregnant, others while they have a joey in the pouch. But none of these copulations are fruitful because the flier produces no eggs to be fertilized until her joey is about 235 days old. Even then, she is likely to reabsorb completely any embryo that starts to develop. Not until her joey is about 283 days of age does her womb become hospitable to a fertilized egg. Its steady growth must not be completed until her pouch is vacant. Generally this program limits the gray flier to one joey each year.

Strangely, the closest relatives of the gray kangaroo follow a different schedule. The red kangaroo *(Macropus rufa)* of the inland plains and the euro *(M. robustus)* of the coastal mountains and rocky central ranges, seem to breed about equally in all seasons. Although they sometimes congregate in mobs of a hundred or more, they also disperse themselves widely when the chronic drought in their habitat becomes especially severe. Prior to the Ice Age they also had to flee native predators, and a flier might easily be a long way off from the nearest male when her time to mate arrived.

Probably in relation to these episodes of separation, the fliers of these two species do not wait as a gray kangaroo does until the joey is about to vacate the pouch before starting another pregnancy. Instead the mother releases an egg and attracts a boomer within a day or so after a newborn baby begins to suck on a nipple. Twenty-four hours after giving birth, she may be pregnant again. Yet the new embryo goes dormant after only about two days of growth. By then it will be about 1/100 inch in diameter, consist of about 85 cells, and be surrounded by a thin shell membrane. It will wait in her womb until the joey in her pouch is about 240 days old (or is lost), and then resume development toward the day of its own birth.

A mother red kangaroo or euro will no longer let her joey into her pouch after 235 days because her next baby is due to be born. Less than a week later she can have her full reproduction line in operation. In her womb there should be an embryo in the 85-cell stage, waiting either until 31 days before her pouch joey is due to leave home or until something happens to it. Her unweaned young-at-foot will be returning to nurse until it is a year old. Her latest offspring will be tethered to another nipple. Two small nipples are left unused, and soon cease to produce milk. She remains unreceptive to a male so long as she has a joey in her pouch, but she can give birth once more without further attention from a boomer. Her next mating will follow the changing of the junior guard.

The programs of kangaroos seem close to the ideal in efficiency. Yet the number that survive the 12 months until they are weaned is rarely more than half that of newborn individuals emerging into the world. Accidents befall a joey even in its mother's pouch, particularly if a young-at-foot is frequently sticking its head in there to nurse. Many of the young-at-foot seem unable to keep up with their mothers and the rest of the mob, and starve without their subsidy of milk. In a panic, a flier will jettison any pouch young that can release the nipple. Panic situations are more frequent now that European settlers have entered the Australian scene, bringing guns, foxes and other predatory mammals. By contrast, the aboriginal tribesmen and the semidomesticated dogs (dingos) they introduced had little effect upon the kangaroos.

The surviving kangaroos sometimes act as though they could still get along with a man. Alan Moorehead tells a charming tale of a young kangaroo that slithered down the bank of a dried-up stream in southern Queensland and sat beside him in the early evening as he watched the flocks of parakeets silhouetted against the sky. The kangaroo was fully aware of Moorehead's presence but unafraid until a noisy Willie wagtail (an Australian bird with the manners of a bluejay!) screamed at the human intruder and sent the marsupial off in panic. Our own experiences with gray kangaroos make us the apprehensive visitors in their midst. We made the mistake of sharing our lunch with a small mob that came close and found that they not only could bound after us faster than we liked to retreat, but also that if our generosity faltered they were willing to lean back on their tails and raise both hind feet against our bodies with a shove that could knock us down. We hope that there will be another opportunity for us to share a grove of gum trees with a mob of gray foresters. But whatever the season, we intend to let them stick to their natural diet while we enjoy ours.

CHAPTER NINE

Finding a Mate in November

THE MAMMALS THAT mate in November, more than those of any other month, reveal an ambivalence in human attitudes toward creatures that combine beauty and utility. Many of them, such as deer, appeal because of their grace. Mountain goats amaze us by their ability to climb mountains. Other mammals, such as she antarctic seals, grow fat in parts of the world that are too harsh and remote for continued human habitation. All of these admired creatures are sought out in their home territories by human hunters for both sport and profit.

The White-Tailed Deer

A few years ago, shortly after dawn, we chanced to see a doe deer and her half-grown fawn walk quietly past our home. They crossed the paved street, followed a flagstone path between the houses of two neighbors, and progressed toward the shrubbery along a narrow stream. Since we were already dressed for our early morning walk, we waited only until the two were out of sight before hurrying out our door and closing it silently behind us. Then we searched with our binoculars, fully expecting to discover which way the deer had gone. Yet their head start of less than a minute was all they needed to safeguard their privacy.

Deer are experts at remaining hidden. Until a square mile of suitable territory holds more than 40 deer, they seldom reveal their presence. Today they fare better than before Europeans reached the New World. Habitat most suited to deer increases every time a person leaves a forest area that has been cleared or abandons a farm, letting it grow up in nutritious weeds and seedling trees. Grazing and browsing according to what they find, the animals slow the pace of change at which the forest re-establishes itself. They respond to good nourishment by maturing early and having large families.

The white-tailed deer *(Odocoileus virginianus)* that the pilgrim fathers met are the only ones east of the Mississippi. Since 1870 these native animals have spread into Nova Scotia and, either by swimming the narrow channel or running across the new bridge, colonized Cape Breton Island. In Canada their range extends from the Gaspé and the environs of Quebec City to southeastern British Columbia. Southward they can be found as far as the Florida Keys and Panama. Thirty different geographic races of white-tails bear identifying names over this enormous range. All of them run away if discovered, raising their snow-white tails like truce flags but depending on leaps and bounds to take them to some new place of concealment.

The tails of the elusive deer tell us more west of Minnesota, Kansas, and in western Texas, where the range of white-tailed deer overlaps with that of the second member of the same genus: the black-tails *(O. hemionus)*. These animals keep separate because of different preferences for foods. Black-tails are somewhat smaller than white-tails, have larger ears, and tend to keep their tails down. Their tails are ropelike and black-tipped on members of the several races known as mule deer, whose range extends from Alaska to Baja California and adjacent parts of northwestern Mexico. At the limit of their eastern distribution, in southwestern Manitoba, they are called "jumping deer" from their habit of leaping over rocks and shrubs while looking back over their shoulders to see if they are being followed. The black-tails with the darkest coats, with tails that are broader, entirely black above, belong to the Columbian race; they are denizens of the humid redwood or spruce-fir forests on the Pacific slope, where the tall vegetation of the understory discourages jumping.

The Columbian deer, like the white-tails, seldom travel any distance according to season. Many of the does and fawns, for example, remain all year within an area of less than 300 acres—not even half a square mile. Bucks wander farther, particularly during the November rut. Mule deer, by contrast, commonly migrate as much as 100 miles every year, from higher elevations in the mountains where they summer to wintering areas in the valleys.

All of the black-tails differ from the white-tails in the pattern shown by the antlers as they develop on the bucks. On a black-tail, the brow tine is followed by two main branches of the beam, each of which fork, and the beam curves only slightly forward as it rises to each side. On a white-tail, every tine rises single-pointed from the beam, and the whole array curves forward over the brow.

No matter where these deer live, the bucks respond to increasing length of days in spring by secreting hormones that induce growth of antlers. Twin mounds appear on their heads just behind and between the ears. Within the mounds cells are dividing and producing the firm tissue that gradually takes on the form the finished antlers are to have. Blood vessels in the hairy skin that covers the firm tissue bring lime and phosphorus according to the quantities of these materials the buck can get from his food. His nourishment, his age, and inherited traits all enter into determining how big the new antlers will be. But until they harden, the buck must restrain his aggressive tendencies so as not to injure the growing cells. It will be October before he can rub off the velvet and polish those head armaments by sham fighting with shrubbery and other yielding objects.

Unlike the elk and deer on other continents, the white-tails and black-tails form no lasting harems. When a buck is ready in November, he sets off to find one mate after another. We sometimes wonder how good his eyesight is, or whether his hormones confuse him, for at this time of year, he abandons much of his normal timidity and seems ready to confront anything that moves quietly. Upon seeing another buck, or a person walking in the woodlands in early morning or late evening, or a doe—no matter how unreceptive—he crouches as though studying the situation and vigorously licks his nose, perhaps to help capture any trace of scent that may come his way. His aggressive antics often end at this point, and no one is the wiser. Or he may decide that the situation is his to master. He elaborately circles the moving object and then rushes toward it, with head down and antler tines aimed.

A smaller buck generally takes one good look at all this armament and runs away. A male that can offer competition is more likely to engage in a shoving match. An unreceptive doe will strike at him with her sharp hoofs, and fend him off. A doe whose reproductive readiness matches his shows her condition by standing still or urinating copiously. We wonder how she can have so much liquid ready, but cannot doubt that its odor contains a message that no buck can fail to comprehend. Certainly his attack turns to attention. Then off the doe runs, leading him a merry chase. If he slows down, as though tired, she circles back to him and tries to lure him on. Sometimes he has stopped following because another receptive doe has come along. Then he has two to enjoy, or sometimes three or four. To maintain his role he will fight off any lesser buck. When a bigger one arrives, he relinquishes his does without a test of strength and trots away to try his luck elsewhere. The does make no attempt to follow him. Their principal show of affection is saved for the buck that mounts them successfully, whom they will nuzzle and lick about the face as soon as he dismounts. Their blandishments count for little. After a quick inspection with his nose to make sure that every doe has been served, he leaves abruptly to arrange another rendezvous.

The sex hormones of the buck drive him on to conquest after conquest, or at least one confrontation after another, until the December days reach their shortest length. Then his contribution to the population ends. By January his antlers drop off, those of the older bucks somewhat earlier than those of the younger ones. Each male abandons the intolerance for other males he has been showing since September. He joins a bachelor club, commonly with two to four members, and keeps company with these others of his sex until his next set of antlers is ready to be freed of velvet.

During the cold months, we often discover well-worn trails in the snow where deer travel in twilight between their daytime hideaways in thickets and the places where they can find tasty browse on shrubs and low branches of taller trees. We may wait in late afternoon to see them pass. Yet in the dim illumination it is almost impossible to tell whether the deer we see are antlerless bucks or groups of does. Generally we assume that if only one or two of the animals are large and presumably mature, while their companions are smaller, the big ones are does and the other fawns of the previous spring or yearlings. By late winter, hunger makes deer more sociable. Feeding in groups of 25 to 30 is frequent. Their trampling in the snow where they can still reach edible branches is often patchy, and earns the common term "deer yard." These are temporary associations, as family groups of a doe and her offspring join in or go off.

The pregnant does reach the end of the gestation period in 28 to 30 weeks, generally in June. Some of them become mothers when just one year old themselves, almost always of single fawns; they were precocious youngsters who attracted a buck's attention at a mere 5 or 6 months of age. Older does have twins more commonly, and some bear triplets. Together these females may increase the deer population by as much as 39 percent in a single season.

In many parts of North America today, the deer herds have increased until starvation in late winter and deaths that can be attributed to malnutrition keep the total number fairly constant. Weakness from lack of suitable food may show in diseases to which a well-fed animal is almost immune, and in a loss of alertness or agility that makes accidents more probable. The actual kill is often made by bobcats, coyotes, or larger predators, which scarcely threaten a healthy deer. Even in well-managed woodlands, the combination of accidents, unforeseen losses, and predation commonly eliminates each year fully 12 percent of the deer that are alive at the end of the rutting season.

In several areas of the continent, the regular mating activities of the deer lead to no obvious increase in the population. This steady state was probably normal enough in former times, when cougars and wolves plus many lesser predators took a share of the fawns and of the aging deer while a shortage of browse in the forest understory caused just enough loss through starvation late each winter to balance the equation between deaths and births. In a few regions, such as the forested parts of Manitoba, the original relation between deer and their habitat continues with little change; the average winter is so severe and long that almost no does give birth before they are two years old; many parents succumb to cold, and losses of young are especially high both in the months prior to birth and in the first month after.

Elsewhere man has reduced the number of native predators and ruled that human hunters are to take their place. Usually this means that a fifth of the fall population of deer must be removed selectively with bullet and arrow before the survivors begin to compete for winter food. Any failure to cull the herd adquately results in subsequent destruction of the browse plants upon which the deer depend. Nor does killing the bucks of trophy size help much. The younger males are quite able to make the rounds and impregnant every receptive doe. Except in carefully managed reserves, the elusive deer succeed so well in their family life that no voluntary program of licensed hunting can adequately control their numbers.

Most people seem unwilling to believe that so many deer can remain invisible. The only animals of this kind they ever see are lashed to some passing automobile toward the end of hunting season. Deer crossing signs mark the highways, but almost never is a deer in sight. Yet game wardens who patrol the woods in spring find abundant evidence of deer that died for lack of food. Usually they outnumber the legal kill. Hunters without licenses often take a large number. Free-running dogs destroy more deer than are struck down on highways by automobiles. For deer-car accidents, the conservation departments keep a tally: almost two per night in western Colorado, mostly between October and April, between 4 and 10 P.M. at marked crossings—more than a third as many as the licensed hunters kill in New York state. It seems incredible in view of the obvious emptiness of the autumn woods after the colorful foliage falls and the animals lose their privacy. Where can the deer be during the day in November? Somehow they keep their secret and their trysts, following a way of life their ancestors evolved long before humankind arrived.

The Bearded Alpinists of the Rocky Mountains

Getting close to a North American mountain goat *(Oreamnos americanus)* is far harder than seeing a little herd of them through binoculars. We'll long remember our first attempt, by a quick scramble through sloping fields of glacier lilies, across snow bridges that threatened to collapse into rushing streams of meltwater, from the road at Logan Pass (elevation 6,664 feet) in Glacier National Park, Montana. To get within camera range meant a climb of about a thousand feet in a quarter of a mile. Haste was essential because the animals were progressing steadily around the high slope toward the shadow of the peak. We lugged our camera but not the heavy tripod, and tried to keep out of sight behind rock outcroppings until the last moment. Apparently we succeeded, for the goats looked surprised when we raised our heads and camera, leaning against the last big concealing boulder. Yet their reaction showed no hint of panic. With the utmost dignity, the several nannies kept the slightly larger number of kids moving in the direction they had all been heading. The only real change was that they no longer paused to graze. One kid even tried to nurse while his mother slowly walked along. Yet our movie record reflects our excitement. It shows the whole mountain ledge and the sedate goats pulsating

Mountain goat (*Oreamnos americanus*)

to the beat of the cameraman's heart, with now and then an extra heave as he gasped for breath while trying to hold steady against the rock!

Mountain goats are strictly denizens of North America, and related to antelopes more closely than to goats elsewhere. Their nearest kin, the chamois of mountains in Europe and Asia Minor, resemble them in possessing soft pads on each foot. These act as suction cups on bare rock where the hard paired parts of a hoof would not hold. Mountain goats of both sexes have horns and similar coats of long shaggy hair and thick underfur. The yellowish white color blends easily with the snowfields where the animals make their home. Unlike the horns of a chamois, which end in a characteristic hook, those of mountain sheep curve only slightly as they taper to the single tip. Only the kids lack a goatee, although that on the he-goat (or billie) is usually more fully developed than that on the she-goat (or nanny). The long hair on the adults, which makes the tail appear bushy and hangs down like baggy trousers over the thigh portions of all four legs, is generally more pronounced on the he-goat.

Most of the kids are born in June while the pregnant females and their young of the preceding year are working their way up the mountain slopes toward summer range. The he-goats are usually ahead, in bachelor groups of three or four, heading for still higher ground. Each mother temporarily stops by herself under an overhanging ledge or in some other shelter where she can give birth. Within minutes her kid is on its feet and nursing. As soon as it is satisfied, she forces it to lie down while she gets something to eat. It leaps up when she returns, and may try jumping about in a stiff-legged way within half an hour after being born. The nanny does her best to stand over her offspring or, with soft bleats, to summon it back to her if it tends to stray. Ordinarily she tends it alone for several days before rejoining the little herd she left temporarily to give birth and introducing her offspring to the other nannies and playful kids. Some of these family groups include yearlings still following their mothers.

At all seasons, the mountain goats prefer the alpine grasses and sedges. The circumpolar tussock sedge *Kobresia* is a favorite, but opportunities to eat clovers of several kinds are rarely missed. To reach these plants, the goats step carefully along trails so narrow that a catastrophic tumble seems imminent. Yet the animals rarely fall or slip, even when buffeted by winds or pelted by hail and snow. If a landslide has sheared away a former route, the goat goes as far as it can, then carefully rises on its hind legs with its belly toward the mountain and twists around to walk back the way it came.

Worsening weather near the peaks and shortening days send the mountain goats toward the lower elevations after autumn equinox. As though frustrated by the increasing scarcity of food where alpine conditions afford sanctuary, the adult goats get irritable with one another. Each he-goat begins threatening his peers, turning broadside toward them while glaring over his shoulder with muzzle slightly lowered; in this posture his horns seem longest and his hairy coat provides the largest silhouette. He pauses at intervals to soil his underside and hindquarters, pawing at the earth and urinating on it, then sitting on his rump and holding himself upright with one front foot while kicking dirt over himself with the other. Sometimes he lies down and wallows. Soon he is conspicuously different in appearance from every she-goat and young animal. Yet all of this is show. He rarely presses a confrontation beyond the stage of an intense threat. Actual fighting with such sharp-tipped horns and where a misstep can carry an opponent downslope to death is too hazardous for the good of the species.

The he-goats roam from one family group of nannies and kids to the next, seeking females that smell attractive. Having found one, a billy stays as close as he can. At first, she will drive him off. His status in the social order is clearly lower than hers. As the behaviorists describe it, "he is subdominant until her agonistic tendencies become depressed." For the day or two until she becomes receptive, he can wait. As soon as she no longer acts as though she would attack him when he comes near her, he tries to mount her. She evades him. He circles and tries again. She runs off; he follows. If she lies down, he stands by for a few minutes and then prods her to her feet, nudging her with his nose or a front leg. All the while, he keeps watch for the approach of any other billy and ignores only those that stay at least 30 feet away. Any closer is too close, and the defender struts over with his flank showing, his head lowered and neck stiffly arched. Usually the intruder retreats promptly.

The rutting urge in November brings the adults together. It also affects a great many of the kids that are no more than five or six months old. They begin mounting one another in sexy play that can have no consequences. It ceases before the goats reach their winter range. There, on land for which humankind temporarily has little use, the goats in larger herds continue their ascetic, dignified existence. Their way of life keeps them out of reach and maintains their numbers more consistently than that of any other native American mammal with an adult weight of 300 pounds.

The Southernmost Seals

The most southerly of seals specialize in living apart. They occupy a circumpolar range all around the coasts of Antarctica, where bitter weather protects them from almost every predatory kind of animal. An English navigator, James Weddell, discovered a few of them in January 1823, and they have been known as the Weddell seal (*Leptonychotes weddelli*) almost ever since. They are large beasts, weighing up to 900 pounds, with small faces and big lustrous brown eyes. They show no sign of apprehension as they watch a person walk toward them on the shelf ice until the intruder is within about three feet. Then the seal reacts, first by emptying its bladder and then by moving clumsily, yelping, to the nearest blowhole or crack in the ice through which it can dive to safety in the cold water.

All winter the Weddell seals live below the ice around Antarctica. Explorers hear the animals calling to one another beneath the frozen roof. They scratch and gnaw holes through which to fill their lungs with air, or they visit natural domes filled with great lens-shaped bubbles. Fortified with oxygen, the seals dive again to gather the mysid shrimps, other crustaceans, squids, and small fishes that comprise their favorite prey.

The search for food and oxygen often takes the Weddell seals far from the actual coast of Antarctica. Then, as the spring days lengthen and the margins of the ice shelf break up, the animals move back south, away from open water. Gradually they approach the mainland. By early October, which corresponds in the Southern Hemisphere to April in the Northern Hemisphere, they may reach latitudes at which a man standing on the ice shelf can see in the distance the mountain peaks of the southern continent. Somehow the seals detect some signal from this site or from the time of year. The cows haul out their bulky bodies, which are as much as eleven feet long, onto the ice and give birth within a day or so. This ends a gestation period that began the previous November or December.

A pup of the Weddell seal is fully four and a half feet long at birth and weighs about 60 pounds. Its thick coat of rusty-gray wool insulates it from the ice, keeping the youngster warm without letting its body heat melt a puddle below its bulky body. The umbilical cord links the newborn pup to its mother for a surprisingly long time. Eventually, she breaks it by swinging her hindquarters from side to side. Then the afterbirth loosens and falls to the ice, and gulls arive to clean up all the debris.

The mother seal fasts for about six weeks while she suckles her pup, and loses about 300 pounds in weight while the pup gains nearly two-thirds of this amount. By the time it is weaned, the youngster may be seven feet long, an experienced swimmer, and already adept at pursuing prey under water. Yet no more than 50 out of each 100 pups may survive these first two months. Some are abandoned by their mothers and soon starve to death. Others get crushed when the bull seals haul out and fight viciously with one another. Still more die of injuries they receive while learning to breathe and feed amid storm-tossed masses of ice as the shelf breaks up.

The slashing of one bull seal at another is particularly hard to understand among animals whose survival is so threatened by violent weather, and whose movements on the ice are so awkward. Unlike the pups, which use their front flippers to move themselves about for a week or two, the adults keep their flippers folded and rely entirely upon clumsy wriggling. Yet their teeth make effective weapons, the second pair of incisors above and below being greatly enlarged, almost as tusklike as the canines. A battling bull aims to use his teeth to bite off his opponent's genitals, and succeeds quite often in doing so or in tearing the other bull so badly that it has no future role in the reproduction of the species.

The victors in these contests on the ice wait as though they expect the cows to join them as soon as the pups are weaned. Instead, both cows and pups leave the ice shelf to feed and do not return. The bulls then dive in fast pursuit. Copulation occurs underwater, after a courtship that only a few men have glimpsed. Thereafter the Weddell seals remain almost solitary until the next breeding season. They stay out of sight and so scattered that no one is sure whether there are 200,000 of them in the world or 500,000.

The bulls that have been maimed in the battles on the ice shelf get away as best they can. No doubt some recover from their wounds. Some swim out under the ice and die; their bodies have been discovered as much as 35 miles from shore, under several feet of pack ice. Others go inland. A few succumb after they have climbed to the surface of glaciers more than 3,000 feet above sea level. Their attempts to escape as far as possible from the bulls that have bested them seem to be extreme examples of a phenomenon noticed among seals of other kinds as well; it is called the "instinct of retirement." Surely it is a strange way to end November!

The Mating of Moby Dick

The whales and their kin, more than any other type of life, have proved to humankind that the seven seas are one. Of the eight different cetaceans in northern waters that the great Swedish physician Carl Linnaeus named in 1758, no less than five swim past New Zealand as well. An additional nine species are almost cosmopolitan in the marine world. The most romantic is certainly the sperm whale or cachalot *(Physeter catodon)*. One individual of special size and sagacity was the white whale that Herman Melville immortalized in his puzzling, allegorical novel *Moby Dick*.

Melville constructed his story around the beliefs of whalers in the middle of the nineteenth century. He gained this lore the hard way, by serving as a seaman on whaling ships sailing from the ports of Europe to the South Seas where whales still swam and courted and produced their young long after those of the northern oceans had been decimated. For nearly 100 years afterward, no scientist had real evidence to support the whalers' claim that sperm whales follow a migration schedule so regular that a knowledgeable navigator could rendezvous with a particular animal according to date, latitude, and longitude. Now thousands of numbered stainless steel darts have been recovered from individual whales that were

Sperm whale *(Physeter catodon)*

made recognizable in this way while still accompanying their mothers as nurslings. The facts confirm most of the conclusions of the sharp-eyed whalers.

The word nursling seems strained when applied to a newborn sperm whale thirteen and a half feet long, weighing 1,700 pounds. Its birthplace is likely to be a calving area to the east of the Fiji islands in the tropical South Pacific, or between Ascension and St. Helena in the South Atlantic. The calf will be on schedule if it is born in February or March. March begins the whaling season, which lasts until October. The whalers show little interest in the calving areas. Good conservation practice prohibits the capture of a nursing mother or her calf. The sperm whale mother suckles her youngster for more than two years, until it is almost 25 feet long. Generally she rolls on her side to let the calf breathe through its blowhole while nursing. Then she everts from narrow streamlined grooves beside her urogenital opening a huge nipple on each side, and pumps milk into her calf's eager mouth.

The warm waters afford almost no food for the cow whales as they bear and tend their calves. During this period of fasting, the adults laze about and live on the fat stores in the thick blubber layer that underlies their thin dark skin. To recover their weight they lead the pod southward as soon as the youngest calves can tolerate the colder water. Already the days are shortening; the drifting plants multiply more slowly. For sea animals of all sizes the amount of energy available is shrinking steadily.

For a while these autumn changes in the Southern Hemisphere do not affect the sperm whales because their principal food is large squids, which they reach by diving to depths of more than 1,000 feet. The squids remain abundant as long as they can prey upon big fishes, which eat smaller animals. Each transfer of energy—from drifting plants to small animals, to larger animals, to fishes, to squids, and to whales—introduces a delay. Their food supply lasts far into the winter as long as the sperm whales stay north of the fortieth parallel of south latitude. (This parallel crosses between the Australian mainland and Tasmania, then the North Island of New Zealand, Chile at the level of Valdivia, and misses the Cape of Good Hope altogether.)

In recent years, the scientists at the Whales Research Institute in Tokyo have paid special attention to sperm whales. Japanese whalers have had to turn to this kind of toothed whale more each whaling season merely to stay in business. They can no longer find enough of the larger and more profitable baleen ("whalebone") whales because these have been exploited to the verge of extinction. Now the scientists are piecing together the evidence and finding fascinating correlations between the age of the sperm whale, as shown by the number of layers of dentine in the 16 to 30 teeth on each side of its narrow lower jaw, and the kinds of whale lice (strange caprellid crustaceans) that creep over its back and feed there. For a dozen years at least, a male calf stays in the pod with his mother, other females, and younger males. He gets to be 37 feet long, which is almost big enough to attract the attention of a commercial whaler. Yet his whale lice are like those of females and younger companions, all of the kind biologists know as *Neocyamus physeteris.*

The twelve-year-old male sperm whale is already bigger than almost any mother sperm whale. Probably his appetite can no longer be satisfied while he is competing with others in the pod. Still he seems reluctant to leave them. Only a few males go off as loners as soon as they reach this age and size. Fully 50 percent stay around until they attain puberty at eighteen years and a length of just under 40 feet.

The males that part company soon encounter in their travels a different kind of whale lice, known as *Cyamus catodontis.* Solitary males of 39-foot length bear both kinds of whale lice in almost equal numbers, which makes scientists wonder why the one kind progressively replaces the other. The family type of whale lice—found on females and young males in the nursery pod—may suffer from the cold twice a year as the male develops his regular migration, feeding far from the equator first in one hemisphere and then the other as the sun brings summer to each in turn. Until puberty, however, the males apparently never invade Antarctic waters. They do go as far north as the Aleutian Islands, Iceland, and Spitzbergen. Almost certainly the first sperm whale to receive a scientific name was a traveling male, for it was caught among the Orkney islands, beyond Scotland at latitude 58° north.

Fully half of the male sperm whales that whalers kill are still immature, no more than 42 feet long. This length corresponds to 24 years of age. The male whale still has one more year and an additional 2 to 3 feet to grow before he loses every one of his family-type whale lice, bears only *Cyamus,* and is sexually mature.

The change makes him a potential harem master. It shows both in the development of his sex organs and in his behavior. His penis, which he keeps curled up inconspicuously most of the time, becomes more than four feet long when erect. His testes, which are concealed just below the skin in front of his anus, grow to weigh about twenty pounds apiece. He develops an acute interest in the pods of female sperm whales and adjusts his migration schedule to visit them. He may detour through the Indian Ocean south of Madagascar, where some family groups of sperm whales congregate at Christmas time—early summer in the Southern Hemisphere.

Rarely do two harem masters visit the same pod simultaneously anymore. Too few of the big male sperm whales remain to compete, perhaps to battle furiously for the right to dominate the cow whales. The arrival of a harem master in a nursery pod occurs anytime between late August and late March, for the big male is ready to mate during months when few females are receptive. He soon moves on if he finds every cow whale still suckling a calf, for she will not yet accept his attentions, and he has no time to waste.

A female sperm whale goes through a four-year cycle after she becomes mature. Her first pregnancy may begin in mid-November, when she is at least 28 feet long. It will end 14½ months later, and be followed by 24½ months of giving milk. During all this time she will avoid a mature male. A total of 39 months is a tremendous investment in the welfare of a single calf. It allows her about nine months to rest and regain her strength before November comes, with another harem master to start her cycle all over again.

The long involvement with each calf makes the sperm whale a particularly vulnerable species. Since few females survive beyond 22 years of age, when they grow to about 36 feet in length, they have an opportunity to bear only a few young, perhaps no more than three. Males sometimes elude the whalers and reach 32 years of age and 60 feet in length. Moby Dick may be credited with being still older and bigger, turned white as sperm whales do at great age. His weight would have exceeded 50 tons.

This monster is an American legend, yet tangible reminders of his kind are on the market. We finger the teeth of sperm whales long dead when we visit shops on Nantucket Island or along the New England coast. We recall that the cylindrical head of these amazing animals conceals a strangely twisted skull, with an S-shaped blowhole near the front, over to one side. Through it the living whale forcibly exhales after every dive, spouting forward a single column of oily steam that can be recognized a mile away.

The sperm oil is discharged in the form of a bubbly emulsion that filled many of the air passages of the whale during its previous dive. The need for this emulsion seems especially great in a sperm whale, since the animal goes deeper than any other wild mammal and thereby subjects its body to enormous pressure. Sperm whales have drowned themselves more than 3,000 feet below the surface by getting entangled in transoceanic telephone cables where the overlying water produces a pressure of about 1,500 pounds to the square inch—100 times ordinary atmospheric pressure. To dive repeatedly in pursuit of squids, the whale draws oil from a huge "case" in its head, and replenishes the supply from a gland. A whaler counts on getting about 30 barrels (1,260 gallons) of sperm oil from the head of a single whale, plus an extra ton of gelatinous spermaceti wax. Additional sperm oil can be extracted from the blubber. Similar oil in the muscles of the sperm whale has always been assumed to make the meat inedible, and most whalers still dump tons of it overboard.

Now the Japanese have found a way to extract the oil from sperm whale meat, and to process the muscle tissue into a protein meal that is suitable at least as a food supplement for domestic animals. A new economic gain from sperm whales makes catching them profitable once more. It may lead to the killing of the big males before they have served for several years as harem masters, and cause sperm whales to follow the various whalebone species into decline toward extinction.

Sperm oil had great importance as a lamp fuel and a lubricant until it was replaced in the 1860s by petroleum products. Spermaceti provided the best raw material for making candles; paraffin became a cheaper substitute. The discovery of Edward Drake's oil well in 1859 created a new industry in Titusville, Pennsylvania, and simultaneously ended the financial gain from catching sperm whales. Sperm oil could not be used for making margarine and other foods for people, as could the oil from the blubber, muscles, and skeleton of whalebone whales.

In Durban, South Africa, we talked with the manager of a whaling factory while filming the dismemberment of a 50-foot whale. It was the second to be flensed on the same platform that day. The man, like others in the industry, showed no apprehension that the supply of whales might end, or that his company promoted the decline. His entire attention focused on finding new markets for whale parts, because former uses kept disappearing. Could the teeth that sperm whales bear in their lower jaws be sold for anything other than curios? What service might we suggest for strips of whalebone from baleen whales, now that spring wire and rubber had replaced them in corsets for women? Which of the many fractions that his chemical laboratory had separated from whale oil would appeal to buyers in industries with which we were familiar?

He showed us a bottle of gelatine made from whale blubber. "Wonderfully pure!" he enthused. "Perhaps Kodak or some other photographic film maker could use it. Whom should I approach with a sample?" His attitude toward exploiting a wild mammal made us think of the statements we read occasionally such as "Due to inflation, the market value of all of the minerals that could be extracted from a human body has risen from less than a dollar to $2.19."

Alive and free, a whale has no measurable value to most people because it can be neither sold nor bought. The instinctual endowment of the great animal, which has evolved to almost crystalline purity in the molecules of DNA that encode it in the chromosomes, is worth nothing when separated from the body and stored in a bottle. The genetic heritage reflects the hazards that whales have overcome for millennia. It fits the present environment, programming the growth of the whale from conception to natural death, but only where the influence of humankind is absent. This heritage spells out the behavior of the nursing mother, of the adolescent male as he departs to feed alone, and of the females to which he may some day return as an eager suitor. It holds importance in the world of the free creature, in a context that began long before anyone recognized November or gave the whale a distinctive name.

CHAPTER TEN

The Year-Enders

WHEN THE FINAL month of the calendar comes into view, most kinds of life north of the tropics have readied themselves for winter. The deciduous woodlands seem lifeless. Ice roofs the ponds and smaller lakes. Bears and groundhogs are in their dens. Squirrels hide in tree holes except near midday. The bats are hung up in hibernation, or have flown south where they can still catch insects active at night. We too can hurry toward the Gulf States for at least a short time, there to reassure ourselves that the stimulating sun is still telling some kinds of mammals that it is mating season.

The Sirens of the Florida Waterways

The largest of North American animals that mate in December live in southern Florida as outliers of a population that mostly haunts the coasts and waterways of northern South America. To associate with them, we go south to Tampa or Miami and sit quietly beside rivers and canals that appear clogged with floating vegetation. Much of it is water hyacinth, which Floridians call "million-dollar weed" because they spend so much to get rid of it even temporarily. In its native South America, it is a favorite food of the mammals we go to meet: the manatees or sea cows (*Trichechus manatus*). There, but not in Florida, they keep it cut.

Manatee (*Trichechus manatus*)

If luck is with us, we see a slate-gray head rise a few inches from the water. A fleshy, two-lobed upper lip grasps a clump of plants and pulls them into the mouth in the broad, bristly muzzle. The head appears to be a foot across, with no narrowing toward the broad rounded shoulders and smooth back. Head and back are all part of one continuous curve that goes out of sight in the dark waters. We cannot see the finlike flippers that are waving gently as the animal maneuvers toward another mouthful. But we need no more of the heavy body in view to know that the manatee has accepted our presence, or is unaware of us.

Only once has one of these incredible creatures passed us where the water was clear and shallow, letting us make out the whole body at once. The shape of the mouth, the tiny eye between fatty folds on each side of the head, the two valvelike nostrils tightly closed, and the semicircular outline to the broad horizontal tail kept us from mistaking it for a small whale. The tail sculled gently up and down, propelling the streamlined body forward at a fair pace about a foot below the surface. Not a ripple or a sound told of the animal's progress.

A male manatee may be thirteen feet long and weigh 1,500 pounds. Females rarely exceed nine feet in length or 900 pounds in weight. Both seem to be harmless, gentle vegetarians, too timid to eat or show themselves when people are in sight. Manatees in Florida are limited in numbers less by suitable food or space than by human activity and occasional frost. Large numbers of them died of cold when the temperature of the water dropped during the big freeze of 1939. Today they cluster in water close to the outflow from the cooling systems of electric power plants in Miami and Fort Myers and where warm springs heat rivers opening into the Gulf of Mexico, most notably the Crystal River near Ocala.

Disturbance seems to be the biggest hazard for manatees. The Russian ecologist P. B. Yurgenson calls it the "trouble factor." Even when the animal is not being pursued, it is likely to respond to the presence of people by ceasing to feed, to rest, and to reproduce. If it has young, it can become too frightened to take care of them, or to regroup them when they straggle off.

Untroubled manatees are often demonstrative, although in slow motion, while they swim close together. Both individuals may raise their heads to the surface and nuzzle, muzzle-to-muzzle, for a minute or more at a time. When several are feeding, they communicate at frequent intervals with underwater squeaky chirps that can be detected with a hydrophone. Some of their sounds are as low-pitched as nine whole notes above middle C on the piano scale; but most are two or three octaves higher.

Naturalists in gradually increasing numbers have been fortunate enough to see a bull manatee pursue a cow of his kind in shallow water. Round and round they go, as though she has no real anxiety to escape but still is unready to accept him. His readiness exceeds hers except for a few unpredictable minutes each year. Then, with ponderous splashing, they mate or she accepts several males in rapid sequence. When copulation ends, the manatees rise to the surface and breathe deeply. Without separating far, a pair generally drift to open water and go to sleep, manatee-fashion, with heads and tails drooping, backs arched upward, buoyed up by their habitat but a foot or two below the surface.

After the mating season in December, the one or two calves develop for about 162 days before being born. By then each calf weighs around 60 pounds. Since it is born in the water, it should not need to be taught to swim. Yet its mother has to raise it to the surface to breathe every three or four minutes for about a week before it learns to follow this routine independently. Unlike its parents, which swim only by tail movements, the calf propels itself with its flippers. It differs also in possessing incisor teeth until its adult dentition grows in. Then it lacks teeth in the front of its mouth for the rest of its life.

A bull manatee usually stays close to any female with young. If a shark or an alligator comes along, he makes a great commotion in the water while the mother and the young hasten off as fast as they can. If her calves are quite small, the mother grasps one under each flipper and swims rapidly. Later the family can reassemble. So far, scientists have not identified an all-clear signal that precedes this aggregation. Probably hearing and sensitivity to underwater vibrations are acute among manatees, compensating for their poor vision and sense of smell.

The folklore that has grown up concerning these animals contains much that scientists have disproved. Apparently the fact that a mother manatee has two nipples on her ample breast led to the invention of the story that she sits up in the water with her head and shoulders exposed to let her calf nurse, and often supports the pink baby with a flipper while it suckles. The truth is that the youngster approaches her from underwater while she is stretched out horizontally, belly down; the calf must hold its breath while it gets its nourishment. For most people it is difficult to credit an old-time sailor with mistaking such a creature for a mermaid, or an old-time scientist with accepting the sailor's story and calling the manatee a "siren." Yet these animals are classified in the mammalian order Sirenia. The fossil evidence leads many zoologists to believe that manatees and elephants had a common ancestor less than 60 million years ago.

Each manatee calf is weaned at about eighteen months of age, but generally it continues to follow its parents for the rest of its second year. At mating time the calf gets out of the way, only to return while the adults snooze. Generally a cow manatee disperses her previous young when the next addition to her family is imminent. How she does this, whether by sound, shove, nip, or some chemical repellent, remains a mystery.

In both Florida and Guiana, the manatees are protected by law and appreciated for their natural effectiveness in devouring the plants that tend to clog waterways and drainage canals. But once these slow-moving mammals get the channel clear, they are in danger of being run down by small power boats or cut by propellers that project below the boats. Probably the animals themselves grow careless when they get used to seeing people who would do them no harm. But the effect is to reduce the population of manatees. Their natural rate of reproduction is too slow to allow for human interference of any kind. A cow does not attain sexual maturity until she is about eight feet long, at three or four years of age. Thereafter she may bear a calf, or less often two, in alternate years until accident or disease strikes her down. The contribution she actually makes to the population of her kind depends upon having a mate, and a quiet place, and her expectation of life. So far no one has discovered how long or short this is, or how fast her productivity is changing as seclusion becomes ever harder to find.

The Big-eared Little Foxes

Proximity to man has already brought disaster to large numbers of the big-eared foxes that formerly hunted rodents and jackrabbits over the American West from British Columbia and Alberta to Texas and Mexico. The first one of them we met walked silently out of the darkness in a forest campground in New Mexico where we had pitched our tent. Suddenly we realized we had company. The little animal, no more than eighteen inches long plus a twelve-inch bushy tail, stood watching us as we ate a late supper close to our gasoline stove by the bright light of a Coleman lantern. Then the fox sat down not more than twenty feet away and studied us with eyes, ears, and nose. Adjusting the position of its head almost constantly, it kept sniffing, seemingly curious about the food odors coming from the pot on the stove and the box of groceries at our elbows. But the oversize ears were the animal's most distinctive feature, aside from its yellowish gray fur. Together, its ears seemed almost as big as its face. Long fine hairs formed a screen across the opening of each ear. We felt sure that they guarded an especially delicate hearing system without interfering much with its sensitivity.

After a few minutes we flicked a piece of cookie to within a foot or less of the fox, trying not to make any extravagant throwing motions with our arms that the animal might interpret as a threat. Instantly the fox was on its feet, half turning to flee. Then, cautiously and gracefully it crept over to the bit of food, sniffed it, licked it, and ate the morsel as though participating in a tea party. Our next contributions fell short. Perhaps the little fox had misgivings about our intentions. Still watching us over its shoulder, the animal trotted off into the darkness and we saw it no more.

Appropriately, America's two kinds of big-eared foxes are named *Vulpes macrotis* and *V. velox*—the "big-eared" and the "swift." Ranchers and desert dwellers know them as "kit foxes" because of their size, and as "swifts" for the speed with which they can race through the darkness, generally to disappear as though by magic—down some old badger burrow or other hole in the earth. We have no idea which of the two kinds of kit foxes paid us the visit, for the range of both overlaps in New Mexico. Their separation is behavioral rather than geographic, for *V. velox* specializes in catching jackrabbits, whereas the preferences of *V. macrotis* run to kangaroo rats and other rodents. When hunting is poor, either kind of kit fox will satisfy its hunger on insects, reptiles, fruits, and even grasses. Unfortunately, kit foxes will also gulp down poisoned bait that has been set out for coyotes or for prairie dogs and other ground squirrels. Human activities have inadvertently destroyed thousands of kit foxes and eliminated them entirely from many parts of their former range.

Few scientists have taken the trouble to get well acquainted with these elusive nocturnal foxes. Harold J. Egoscue made himself an exception when he undertook to keep track of those living west of Utah Lake in Tooele County, Utah, in an area of about 25 square miles. On this tract of desert and semiarid land, he found four or five mated pairs and enough extra adults living alone to conclude that a mature kit fox needs about two square miles of territory on the average merely to stay well nourished. To patrol 1,280 acres takes a lot of running. To meet another kit fox in such a sparse population takes big ears that can pick out the voice of another individual far away.

Probably the vixen (the female fox) forms the pivot in each territory. One in Tooele County maintained her residence in the same small area continuously for four years. Between mid-December and the end of January, she attracted a male who stayed with her until September—all through her six- to seven-week pregnancy, another ten weeks of suckling her four to seven pups, and until the youngsters were ready to go off on their own. Three other families, each at a good distance away, consisted of one male and two vixens, each with a litter of pups. Yet bigamy in kit foxes can be no sinecure for the male. He works just as hard as his mates in bringing fresh food to the den for his offspring. All of the parents lose weight between June and August, when these family responsibilities demand so much of their energy. Their success can be measured both in the number of pups that they contribute to the population, and also in the assortment of bones that are scattered about near the den, discarded remains of prey the parents have hauled home.

The Raccoon's Tropical Cousins

From the northern border of Mexico to the southern limits of Brazil, two hyperactive mammals—the coati *(Nasua nasua)* and the kinkajou *(Potos flavus*)—show how much versatility has evolved among members of the family to which the familiar raccoon belongs. Both of these tropical animals are native to Panama, and make periodic visits to the laboratory area on Barro Colorado Island in Gatun Lake, within sight of the ship lanes of the Panama Canal. There they distract sober scientists by showing insatiable curiosity and amazing ingenuity in reaching food.

The distinguished ornithologist Frank Chapman, who helped found the research facility on the island, began testing coatis by suspending ripe bananas one after another on three-foot lengths of string from an endless steel clothesline that overhung a valley between a pulley on his cabin and another on a jungle tree. Each day a coati would go up the tree and clamber along the clothesline as though it were a squirrel or a monkey, instead of a creature built more like a long-nosed cat. The coati's paws, of course, resemble in their flexibility those of a raccoon. With them the animal clung desperately to the clothesline, trying at the same time to keep its balance by wild attempts to use its long tail like a tightrope artist's horizontal pole. But eventually, after many near calamities, it would reach the string supporting the first banana. There it would pause, momentarily frustrated by the length of string between the clothesline and the fruit. After a few failures in reaching for the banana while clinging with hind feet alone, the coati would begin using its forepaws systematically to haul up the prize. With many a slip, each animal persisted until it could take the banana firmly in its mouth. But now, how to enjoy the reward? Could a coati eat its banana while clinging to the clothesline? Or would the acrobat attempt to make a return trip to the tree while holding its uneaten prize in its mouth?

Different coatis showed their individuality in their solutions to these tests. Many a banana dropped to the valley below whether the animal chose to eat on the wire or to go back with it to the tree. But one big coati became the master of the situation. It learned to pull up the first banana, eat it with only the loss of minor crumbs, discard the skin, and progress to the next banana—until its appetite could no longer impel it to further feats. For years this particularly skillful animal returned and repeated its performance daily whenever Frank Chapman spent a while on the island. Seemingly its skill on the clothesline was matched by other outstanding abilities that saved it from the normal hazards of coati life.

The name coati (or coatimundi) came via the Spanish and Portuguese explorers from the descriptive phrase used by the Tupi Indians of the Amazon Valley to identify the animal. It is "the one who sleeps with his nose at his belt," for the coati does curl up in this curious position, wrapping its long tail around its body. The tail may be 27 inches long, compared to only 26 inches for the rest of the coati. The appendage is banded across the top in a pattern that reminds us of a raccoon. But ordinarily the animal holds its tail vertical, with only the tip curled over.

Coatimundi (*Nasua nasua*)

Often we have encountered a little group of six or eight of various ages coming along a trail in the rain forest. First the tails came into view, and then the coatis themselves, their noses down at ground level, sniffing for whatever food they can find.

In Central America, the common names for these animals differ according to behavior, which varies according to age and sex. An old male tends to hunt by himself, and is a *pisote solo.* (*Pisote* is a corruption of the Nahuatl, Aztec, name *pitzotl.*) Females and young, who hunt together, are *pisotes de manada* ("herd coatis"). We have often wondered whether still another name is needed in December, when the male becomes sociable and the females chase away their young companions to give attention to a mate!

Quite possibly the courtship is too brief to affect the names that people use for coatis. No one seems to know whether the male stays near his mate for more than a few minutes before going off to find other receptive females. The mated female has about eleven weeks of pregnancy before her four or five young are born. Within a few days they are following their mother everywhere, to the water's edge in a stream or lake, up over fallen trees, and through the undergrowth where tropical sunlight can reach the ground and make low plants flourish.

Apparently a young coati is expected to keep up with its mother and the rest of her family. Any that lag behind are likely to be pounced upon before the parent can rush to the defense. Yet some do lag. We discovered a baby coati at the foot of a big tree, all by itself, as though it had been left there and told to stay. When no mother coati returned for the youngster while we watched it for five minutes or so, we bent over and picked it up. At this affront to its dignity the baby let out a shrill scream, more like a bird call than any sound we had previously heard a mammal make. Less than two minutes later, an adult coati appeared, looking about in a most agitated way. We set the baby down, and it ran to the big coati—evidently its mother. In a matter of seconds, the two vanished together down the trail and into the undergrowth where, we assume, the rest of the little family was waiting. Now we wonder why the baby coati did not summon help before we picked it up.

After a young coati is weaned, its dietary possibilities seem endless. In a tropical rain forest, it has about it a veritable arboretum loaded with lesser plants of myriad kinds. Different fruits drop to the ground week after week, and among these the coatis choose the edible kinds. Insects and the insectivorous reptiles are not quite so varied and much less numerous, but the coatis eat them almost indiscriminately. Often a coati climbs a tree to stalk the larger foliage-eating lizards called iguanas, which sunbathe on horizontal limbs. Along the way the omnivorous and perpetually hungry explorer may enjoy a few eggs from a bird's nest, or the nestlings if some are available.

We prefer the eating habits of the smaller relative of the coati and raccoon—the kinkajou or honeybear. Its food is mostly juicy fruits, which it picks with its front paws and carries to its mouth, usually while high in a tree and holding on with both its back feet and its long prehensile tail. The short nose of a kinkajou seems to accentuate its long and active tongue, which the animal uses to get every sweet drop from the fruit it eats and, at intervals, to lick its lips enthusiastically and wash its face to keep it from getting sticky.

The short, thick fur of a kinkajou comes in many shades from pale buff to blackish brown, but seldom meets the light because these animals prefer to sleep all day, curled up in some tree hole or other shelter. At night they explore high in the trees of the rain forest, sometimes alone but more commonly in pairs. Probably they return to some favorite tree before dawn each day, perhaps the one in which the female has a hideaway in which to bear her young. She leaves them at home until they are about seven weeks old, by which time they can learn to hang by their tails and feast on fruit beside her.

For most of its life, a kinkajou is high above the ground. Its activities at night take it from tree to tree by aerial highways that intersect where the outstretched branches from one tall trunk brush against those from another. With a long strong tail as a fifth hand, the kinkajou can move securely, deliberately, and with reasonable speed. Sometimes it will join two dozen of its kind in feasting where one particular tree is loaded with a favorite fruit. It is then that modest controversies break out among the gourmet kinkajous, with grunting sounds and some chattering. In the darkness it is hard for us to see all that is going on. But at intervals our ears can pick out a distinctive call that Dr. Walter W. Dalquest of Louisiana State University describes as "a rather shrill, quavering scream that may be heard for nearly a mile."

A kinkajou on the ground is silent and wary as it moves across a clearing from one area of forest to another. Its legs are scarcely longer than those of a dachshund. Caution is preferable to attempted flight in escaping from a predator. Nor does the kinkajou offer any impressive defense when cornered. We easily caught one that was crossing a road in twilight. The animal quieted down in just a few minutes and accepted a gift of fruit, sitting up on its haunches and smacking its lips loudly as it munched away. It clung to the fruit even when we took hold of its tail and lifted it, inverted, from the ground. The kinkajou merely curled the tail tip around a human wrist as a substitute for a tree limb and continued its free meal.

With such a wealth of wild food available and the cycles of rain and clear weather so repetitious day after day in the tropical rain forest, it seems strange that kinkajous should wait until the end of the calendar year to feel the urge to copulate. The innate choice of December, after eleven weeks of pregnancy and ten more of suckling, points automatically to May as the time for young kinkajous to find wild foods in place of milk. However, this timing matches the beginning of the rainy season in only part of the range that kinkajous occupy. What else might matter? The month when the juveniles go off on their own must be critical, a time of near crisis. When a young kinkajou finally leaves its parents, it needs in its favor every predictable amenity of the environment if it is to find a place to live and raise a family. Presumably the number of places for kinkajous is as stable as the rain forest itself. Any newly independent individual is most likely to succeed where, for some temporary reason, the previous population has diminished slightly. This shrinkage may well be seasonal and may be the ultimate benefit that the parents confer by timing their matings at the end of the year.

CHAPTER ELEVEN

New Year, New Beginnings

SINCE ROMAN TIMES, the Christian year has begun with a month named to honor a pagan god (Janus) with two faces—one looking backward, the other forward. No earlier culture (and none that began in other lands of the Northern Hemisphere) showed such a special joy over seeing the sun stay above the horizon longer every day. A fresh start in the endless cycling of the seasons is now celebrated worldwide by people who follow this cultural tradition. Seldom do they realize that familiar neighbors among the mammals have long been celebrating January in a different way—by mating instead of making New Year's resolutions.

Among those animals that react regularly a few weeks after winter solstice are a good many that humankind has maltreated. They have made a comeback in recent years by investing in new generations according to an inherited schedule established millions of years ago. Some of the squirrels have responded to the removal of a bounty offered for their tails. The beaver outlived the fashion for fur hats. The gray whales along the California coast renewed their populations while warships instead of whalers patrolled the North Pacific during World War II. A change in range let coyotes and bobcats outwit the men who sought to exterminate them. Even the great jaguar, a Latin American leopard, seems to have held its own by making full use of every January.

Big Bushy-Tailed Tree Squirrels

All of the gray squirrels that one of us met for many years were jet black; naturally, they were known as black squirrels. This is the kind common around the western end of Lake Ontario, from the Toronto region to Buffalo, New York. Rumor there holds that the blacks persecute any gray squirrel that appears. Blacks, which are of the same race and species as grays merely are a mutant strain with extra pigment in hairs and skin.

Mutant blacks and pigmentless albinos turn up in many places over the wide range of the eastern gray squirrel *(Sciurus carolinensis pennsylvanicus)*, which extends from the hundredth parallel of longitude at the Gulf of Mexico to the southeasternmost corner of Manitoba and the Atlantic coast north to the Canadian border. A generally similar gray squirrel *(S. griseus)* with a distinctly broader tail lives in woodlands from Puget Sound to southern California. Both are graceful animals, most active in early morning and late afternoon. They are abroad on fine days in every season, mostly in the trees unless they can find food such as nuts, seeds, small fruits, and edible fungi by hopping about over the ground. The gray squirrels that visit the bird feeders outside our New England home always discover the scarlet fruits of the flowering dogwood trees after the robins and starlings have plucked off almost every one; the squirrels creep out on the branches to inspect the fruits the birds have left, and pick over those that have fallen to the ground. We notice that only a few are eaten. The remainder, twice discarded, prove almost invariably to be malformed inside. Yet nothing is wrong with the acorns that the squirrels bury in small holes, repacking the earth while looking around as though to memorize the location. Dozens of acorns sprout before they are found again. In colonial times, the squirrels helped in the same way to plant the fruits of beech, hickory, and American chestnut trees, renewing the forest that was their home.

Some of our neighbors still take action when gray squirrels descend from the trees to enjoy ripe ears of Indian corn, or to dig up imported tulip bulbs and other treasures from suburban gardens. But today few housewives make squirrel pie, or long for a fur coat made of soft squirrel pelts. Far more people put out food that encourages the squirrels to stay around. If nothing else, the squirrels give dogs and cats some exercise until the pets get too old to pursue or stalk any creature so alert and nimble. Sometimes we see a squirrel playfully risking its life to tease a cat, only to scamper up a tree at the last moment. The squirrel may hang head down by twisting its hind feet to hook strong claws into the bark, or it may perch on a low limb, looking down and seemingly jeering at its frustrated pursuer. The clucking calls, ending in a bleated *a-a-a-a,* are accompanied by a vigorous flicking of the bushy tail. Even if the cat decides to climb after

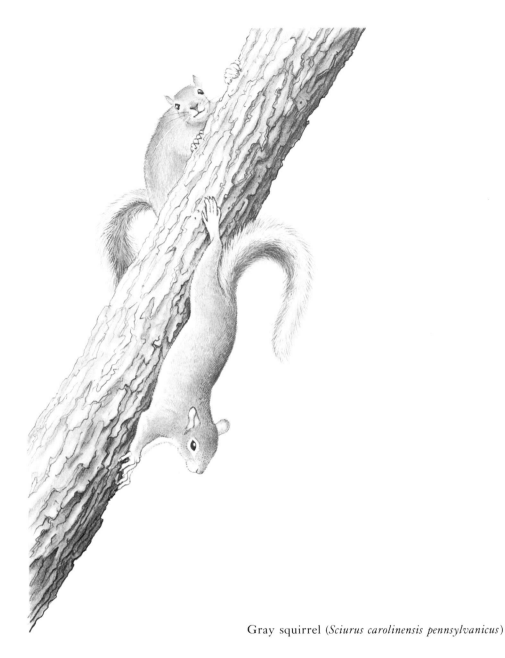

Gray squirrel (*Sciurus carolinensis pennsylvanicus*)

the squirrel, it is in no danger. The squirrel can retreat to branches too flimsy to support a cat, or sky-dive to the ground and run away unhurt while the cat backs down the tree.

The words squirrel and *Sciurus* are both derived from roots meaning "shading tail," as though the appendage served more as a parasol than a signal, a rudder for sky-diving, a counterbalance in walking a slender support, and a blanket in which to wrap the body in cold weather. The long hairs on its tail are replaced once each year, not twice as are those on the rest of the skin. These molts come logically in spring and fall, whereas the inner schedule of the animal calls for sexual activity in winter and summer.

The dizzy chase that a female gray squirrel leads can be followed more easily in January than in July because the deciduous trees stand bare of leaves. From one swaying branch to the next she goes, like a trapeze artist with no safety net below. The irregular route takes a spiral course down a big tree, generally just to double back up again. Rarely does she scoot across the ground to reach another trunk. Her attention may be too distracted by her efforts to evade the male for her to look out for foxes, coyotes, bobcats, or other terrestrial predators. In a tree, she is more likely to notice the approach of a hungry hawk and to dodge to safety. The bird will be moving, silhouetted against the sky, whereas the beasts of prey so often stand still, crouched in readiness to pounce.

The scratching of claws against the bark is often the only sound we hear while a male squirrel is chasing a potential mate. Quite possibly he and she are conversing in ultrasonic frequencies to which we are not sensitive. When he tires he may tell her so and request a breathing spell. Often both squirrels stop, flat against the same tree a few feet apart, perhaps one halfway around the trunk from the other and thereby out of sight. Then off they go again. We rarely see the conclusion after watching the chase, for it may continue through several acres of woodland. But somewhere the game ends in a predictable way for, by February, almost every female squirrel more than eighteen months old is pregnant.

Six and a half weeks is a fairly short gestation. It gives the mother ample time to select a tree hole for a nursery, or perhaps two or three different ones not too far apart in the event that she is disturbed in one and must move her helpless young to a new home. Generally she finds time to arrange in each hole some dry leaves and soft fibers from various plants in the form of a mattress. Her newborn young will need this insulation in the winter, for their one-ounce bodies are naked at first. If only two are born, rather than the usual four or five, they have a special need to conserve body heat because two in a huddle expose so much more surface area per baby than when each one snuggles up to a larger number.

All aspects of parental care are the mother's responsibility, for the father squirrel has found too many mates to extend his attention to their families. The mother spends every night, and midday siesta time as well, with her young, warming them, suckling them, affectionately keeping them clean. If she must, she will carry them one at a time to another nest, each youngster with its legs and tail wrapped around her neck while she goes from tree to tree. She will wean them when they are six to seven weeks old and let them explore outside the nursery at eight to nine weeks of age. Her first litter for the year emerge where we can observe them in May, and her second in October or early November.

By July and January we see three sizes of gray squirrels. The smallest have had only two months to explore the outdoor world and escape its dangers; these individuals are too immature to show any interest in sex. Middle-sized squirrels are six months older, about half grown, and amorously inclined. They practice chasing one another in much the same way the adults do, with shorter pursuits and considerable copulation; quite a number become pregnant; often their matings are not yet on schedule, and produce young in months between those in which the mature mothers give birth. The half-grown young have not yet begun to interest or challenge the mature males; the families they produce are generally quite small.

Obvious to an adult gray squirrel but not to a person is a social hierarchy among the full-grown males and females—those two years old or more. It is an interaction based upon threats and occasional vicious scuffles with which the members of both sexes divide up territory and the perquisites of parenthood. In general, a male is dominant over a female of the same age, and an older squirrel over a younger one. A male tests his social position, just as he hunts for food and mates, over a broader area than a female. Even among the juveniles less than a year old, the average size of the home range is 5 acres for a male and less than 4 for a female. Among yearlings, these dimensions increase to almost 11 acres and just over 7, respectively. An adult male will attempt to dominate all other males and females in about 24 acres of woodland, whereas the mates he finds will seldom be familiar with as much as a 10-acre tract, and around the borders their territories will overlap with those of other mature females. In 100 acres of good deciduous forest, there may be only 4 dominant males, but more than 12 females with litters twice a year. The 4 males will have fathered most of the young. Counting the nestlings, the juveniles, the yearlings, and the adults, the total number of squirrels in the 100 acres is likely to be 230 or more. It may rise to 1,800 after a series of years in which the trees shed an unusually large number of acorns and other fruits that squirrels can find all winter.

The ups and downs in the population of eastern gray squirrels follow a cycle with peaks every five years or so. Interactions among the individuals and competition for nest sites grow so severe that hundreds or thousands of the younger and subdominant squirrels begin cross-country treks. They seek a better home, but generally perish in the attempt. Like the fabled lemmings, some drown while trying to swim across rivers and lakes. Many get run over on the highways. Great numbers that started out fat and well fed find too little to eat along the way; they die of hunger and exhaustion. Hunters report finding the dead squirrels, or the live ones that are moving through the woods, singly but in some definite direction.

The cyclic changes in the populations of eastern gray squirrels reflect the reaction of the individual animals to crowding. It is an automatic response that prevents increases in numbers beyond a tolerable density. It prevents the squirrels from adversely affecting their habitat, and is quite different from the reduction in squirrel populations that follows felling of the forest. In America at the time of the Civil War, the homesteaders destroyed the habitat for squirrels in most states east of the Mississippi by clearing the land with axe and fire. After uncontrolled erosion washed away the forest soil from rolling hills, the farmers moved west to try their hand elsewhere. By World War I, when conservationists began to wonder whether the squirrels had any future in eastern states, tree seedlings had already begun to sprout on abandoned farms. The land grew up in woody vegetation among which squirrels could find food and shelter. Reproduction of the bushy-tailed rodents kept up with their opportunities.

No one seems quite sure whether disease or intensive hunting during World War I, when meat in Europe was especially scarce, led to the extensive disappearance of the native tree squirrels *(S. vulgaris)* that are the Eurasian counterpart of the North American grays. These squirrels are almost identical in size, more reddish or dark gray than the American species, and have conspicuous tufts of black hairs on their ears. In any case, European gamekeepers decided to improve hunting by introducing the eastern American gray squirrel. It proved more elusive and enterprising than anyone had expected, and soon began to reoccupy the forests from which the native squirrels had been eliminated. Then it went on to compete with the surviving Eurasian tree squirrels. Without any real battles, the newcomer took over. Yet when we compare the family life of the two competitors, it is difficult to understand the reason. The Eurasian species takes the same food, uses the identical nest sites, and moves about at the same time of day in a remarkably similar way. It breeds when one year old, has slightly larger families, and sometimes produces five litters in a year by mating at almost any time from spring through late summer. If anything, it would seem to have all the advantages.

The gray squirrel also thrives in eastern North America, where it faces a greater variety of competitors. In thickets and open woodlands from South Dakota and Rhode Island to the tip of Florida, it encounters the larger and slow-moving fox squirrel *(S. niger)*. In the northern coniferous forests as far south as Illinois, and in the Appalachian mountains to Kentucky and North Carolina, it runs from the smaller, more agile and excitable red squirrel *(Tamiasciurus hudsonicus)*, a rodent without counterpart in the Old World. Curiously, both the fox squirrel and the red squirrel possess at maturity one less premolar tooth on each side in the upper jaw than most squirrels, including the grays.

Fox squirrels come in many colors in addition to foxy. Those in the Southeast are responsible for the name *niger* (black), for they are so dark as to appear black except for a yellowish white nose and ears. To the north, along the middle Atlantic coast, are steel-gray fox squirrels. Those in the Mississippi basin are the brightest, orange or brownish orange along the back, shoulders, rump, and sides of the tail, but grayish below except on the tail where glistening white hairs form a lengthwise stripe below and above. Elsewhere the majority of fox squirrels are sand-colored. All of these animals have a squarish face. They grow to be ten to fifteen inches long, with a nine- to fourteen-inch tail, as compared to the ten-inch maximum for both body and tail of the ordinary gray squirrel. In late autumn, a fox squirrel may weigh three pounds, which is twice as much as a gray.

In keeping with its greater weight, the fox squirrel spends more time on the ground, moving slowly with a rolling gait that is almost a waddle. If approached it may bound away among the undergrowth rather than climb a tree. It refuses to go far, however, for its territory is rarely more than 300 yards across.

If we see a fox squirrel before noon, it is almost invariably flattened out lengthwise atop a sturdy limb high on a tree, taking a sunbath. Apparently it rises late, and is not willing to end each day until the light fades after sunset and the woodland becomes too shadowy for diurnal squirrels. Most of the time the animal stays silent and relies upon lack of motion to let it blend with the bark on a tree trunk or become a mere bump on a horizontal limb. Or it creeps into a knothole den or a leafy shelter where it is completely out of sight. So many of these squirrels have been shot at that they tend to be secretive if they see a person in the woods. Only the most daring of them will wave its bushy tail and repeat a quacking call that sounds more like a bird than any mammal.

The most daring are the males in January and July, when their sex hormones are urging them to explore for mates. It is then that pairs of these big squirrels go racing through the forest, taking chances on branches that barely sustain their weight. Quite often the twigs do break off. We wonder whether these breaks on maple trees are the places to which the fox squirrels return in March to sip sap where it is oozing from the wound. They are inveterate tipplers, and sometimes extend the damage by stripping off bark and eating the soft cambium tissue where new growth is beginning. Probably by spring the squirrels have eaten most of the nuts and fruits that they cached the previous autumn, and they get some variety by nibbling the buds on oak, elm, basswood, and black birch.

In their family life, the fox squirrels follow a program almost identical with that of gray squirrels. Yet the hazards they face must differ, perhaps because of their larger size and slower movements, because the populations of fox squirrels show no pronounced change on a cyclic schedule. Like the western grays, the fox squirrels fit inconspicuously into the forest environment. They sustain themselves and help plant the seeds for new trees without signs of crowding or of mass emigrations.

The Dam Builders and Canal Diggers

For changing the forest community, rather than maintaining it, the beavers of North America and Eurasia are second in effectiveness only to man. They fell trees from one to twenty inches in diameter and haul them away, generally piecemeal. They tunnel in the banks of broad, deep streams, hastening erosion. They build impressive dams where they find river water flowing more quickly. For a while the pond behind the beaver dam gives a place for fishes and other aquatic animals to live. Gradually, however, silt and organic matter settle, making the pond too shallow to serve beaver needs. The dam builders move on. The untended dam collapses, releasing the water and exposing an impressive thickness of soft new soil as a seed bed for a level meadow. Trees colonize the meadow and renew the forest. The stream cuts through the soil to its former rocky bed, and once more runs quickly. Its banks erode during every flood. The site again becomes suitable for a beaver family to build a dam and create a new pond.

By choosing particular trees to fell, a beaver gets food and building materials at the same time. Its preference is for quick-growing trees that thrive in moist soil: poplars (including aspen and cottonwood), red (or swamp) maple, alder, ash, black birch, and willows. All of these offer edible leaves in season, tasty buds, tender twigs, and a nutritious inner bark. To get them, the beaver sits back on its hind legs and flat tail close to the tree and gnaws a deep open groove all the way around the trunk. Just before the tree topples, the sound of snapping fibers warns the beaver to get out of the way. Since it ordinarily fells trees that are close to the water first, and these commonly lean away from shore, it gets credited with engineering the direction of fall when actually the animal is saving itself work. As it cuts trees progressively farther from the water, the direction of fall may continue to suggest intelligent action, although the true explanation is more often that branches of untouched trees resist the toppling of the cut trunk in any other direction. A far better measure of the innate engineering sense of the beaver is a long straight canal across level land, dug by the animal from its felled tree to the pond. After removing the branches, the beaver can drag sections of the trunk into the canal and float them to deeper water. Some of these canals are more than 700 feet long, 2 feet deep, and 3 feet wide.

During the summer and early autumn, the beavers seldom cut trees because plenty of nonwoody plants are available as preferred foods and only an occasional peeled pole or two is needed to reinforce any break that forms in the dam. Small branches and mud, perhaps with a stone or two, make better patches. Later in autumn the program changes. Then comes the time for beavers to store away their winter food in a community pantry located in the deepest part of the pond or river. Instead of canned goods, the animals hoard branches and lengths of tree trunk with the bark still on, either stuck firmly into the bottom or held down by stones. To reach this nourishment the beavers must always be able to swim below any roof of ice from the underwater doorway of the lodge they have built in the pond or the den they have excavated in the river bank.

By fall the average beaver more than two years old is so fat that its hips seem especially broad; it weighs 25 or 30 pounds. Among the rodents of the world, it is second in size only to the South American capybara, which grows twice as big. Merely to keep from losing weight, a beaver of these dimensions needs to eat the nourishing parts from a poplar tree one to three inches in diameter every day. Yet if only a pair of young adults are occupying the den like newlyweds, they rarely provide themselves with enough food in cold wet storage to meet all their needs through a four-month winter. If they are more experienced animals, accompanied by two or three of their half-grown kits from the previous litter, their reserve of bark and soggy foliage must be shared among more mouths and usually is still less adequate. To survive until spring the adults lose weight while their young grow slowly. All members of the family huddle together to conserve heat and minimize the muscular exertion that would increase their hunger above the bare minimum.

The supply of winter food limits the distribution of beavers locally and continentally more than any other natural factor. Wherever beavers can no longer find suitable trees to fell within a reasonable distance of the water, they must either move elsewhere or run extra risks from predators that are likely to intercept them on land. The same predators kill many beavers, young and old, that are trying to find a new home. Beaver meat is said to be delicious for human use as well, but the thick, lustrous, waterproof fur that keeps beavers warm in winter has held still greater attraction—particularly among Europeans. The Eurasian beaver *(Castor fiber)* was extirpated in Britain, probably by 1200 A.D., but managed to survive in Scandinavia, along the Rhone, and in some other regions across central Europe to western Asia. It has become almost exclusively nocturnal and secretive, building few dams or lodges that betray its presence.

Until the European market for beaver-fur collars and hats collapsed through a change in fashions, the search for North American beavers *(C. canadensis)* led trappers to explore most rivers and lakes in the area that has become the United States and Canada. Beaver pelts became a medium of exchange and the principal trade item for which the Hudson Bay Company was organized. Trapping became so intense that these animals vanished from the eastern states and the more accessible parts of the provinces. Now they have recolonized, often with human assistance, and can be found in Canada from Newfoundland and the Maritimes to Hudson Bay, Vancouver Island, the Yukon, and into southern Alaska. In the continental United States, they are missing only from Florida and from most of South Carolina, Georgia, Louisiana, and Nevada.

In all of these areas of the New World and the Old, beavers mate inside their dens during January and February. From scientists who have eavesdropped with microphones installed in beaver lodges we know that the animals grunt and chuckle, squeal a little, and scratch a lot, whereas outside their dens they are almost invariably silent. Except for the sound of teeth cutting into a tree, or of a tree being dragged, a beaver rarely reveals its presence outdoors by sound unless it slaps its tail against the water as an alarm signal—warning every beaver in the vicinity to dive for safety.

Apparently the adult male slips away from his mate at breeding time and seeks the favors of any other adult female he can find. He may cautiously enter a den that is strange to him and be ejected, torn and bleeding, by the resident male. Or he may chance to arrive while the other male is also away, and enjoy a temporary liaison, departing without being detected and attacked. We naturally wonder whether male number two is mating with the home female of male number one while this is going on.

About 90 days after copulation, a mature female gives birth to from two to eight kits, each one fully furred and with its eyes open. The mothers that are just two years old and bearing their first babies have the smallest families. Later the number at a birth rises to average between four and five. More of the embryos that begin developing continue through to full term instead of being absorbed by the mother. Probably this indicates merely that older mothers have essentially completed their own growth and have had a chance to fatten up; they are able to share nourishment with more offspring than a first-timer can.

A mother beaver supplies milk to her youngsters through four big nipples, two on her chest and two lower down. The second pair dry up first, beginning soon after the nurslings are four weeks old. By then she is bringing home soft plants from around the edge of the pond and starting the weaning operation. Ordinarily the kits are confirmed vegetarians after six weeks of age, but they stay with their parents through two more winters while they learn by experience and observation. Their mother usually evicts them about a week before she is due to give birth to a new litter, and hence a week before their second birthday. Out they go, never to return. They explore upstream or down, or cross into another watershed if they can reach it without being killed. Some of these young are already pregnant, and all of them will be sexually mature before January comes again.

In the Far North, the young beavers customarily remain with the family for an extra year. Through parts of the Canadian Northwest Territories, the Yukon and perhaps Alaska too, the average family during midwinter includes the parents, the half-grown kits of the previous spring, and full-grown youngsters from two previous years as well. Part of the winter, at least, is spent in dormancy. Possibly the mother fails to recognize that some of her family is mature. More likely the extra bodies conserve heat more effectively as they huddle in the lodge through a long period of cold weather. Whatever the reason, it is three-year-olds that go off to try their luck at colonizing other areas. Already they have spent a larger proportion of their lifespan in family interactions and learning than most animals that survive in the wild for no more than fifteen to twenty years. Possibly the northernmost beavers learn more slowly, or have need for extra training.

Our own experience shows that, wherever beavers are undisturbed by people for several years, they increase their activities in broad daylight. They learn quickly to return to any shore area where apples or carrots are left out for them. Soon they will wait in plain sight while the person who brings these treats renews the supply. A two-year-old or even a parent may crawl out of the water and approach a motionless benefactor, perhaps accepting another gift from an extended hand. In a few days more these bulky wild animals gain so much confidence that they will rest their forepaws on the thigh of a kneeling person the better to reach a handout. If several beavers are receiving goodies at the same time, they give way to one another most amiably, with never a sign of conflict. Their big orange incisors could be formidable weapons, but seem used only for food and cutting trees. A quick return to the water is the beaver's reaction to any danger. By making no gesture or sound that might threaten them, we have been able to stroke a beaver's fur gently without alarming the animal or affecting the rhythm of its munching on a snack.

Since 1971 a family of beavers has consolidated its claim to the millpond and surrounding land in the New England community we too regard as home. On autumn nights the beavers explore much of the acre we own along the pond edge, and take trees around which we have put no protective shields. Wet runways down the slope to the water show where the animals have made repeated trips to haul their branches and lengths of trunk toward storage for winter food. They leave foot-long projecting stumps with pointed tips, which, for human safety, we tidy up with a handsaw. Many trees, of course, regrow from the stump, only to attract a beaver a few decades later. Dozens of irregular boles on our acre prove that beavers have been following these same traditions for uncounted years. Even the sturdy oaks contribute their share to keeping the bark-eaters thriving from one January to the next.

On a sunny New Year's day the big beaver lodge is conspicuous amid the leafless alders on a little island in the mill pond. Snow generally covers the beaver's home and does not melt on the north slope, which receives no direct sunlight. The insulated roof of mud and branches keeps in the body heat of the quiet animals deep inside. Their goings and comings beneath the ice, to retrieve food from their submerged supply and haul it through their underwater doorway, proceed unseen. By working so diligently ahead of time, the beavers have isolated themselves from our wintery world as completely as though they had flown to Florida.

The Latin American Leopard

We would not expect the most experienced beaver to survive long away from home if it encountered a jaguar, which is the heaviest and most powerful cat in the New World. Fortunately for beavers, the ranges of the two species overlap only in Texas and across New Mexico to eastern Arizona, all within 50 miles of the Mexican border. Jaguars show no hesitation in pursuing their potential prey into water, through dense brush, up trees, over open country, or up craggy cliffs. On a few occasions, jaguars have attacked people without obvious provocation. On others, these cats have moved off without a fatal confrontation. Our own preference would be not to surprise so dangerous and unpredictable an animal, for we always go unarmed.

In at least one tribal language in South America, the name *jaguara* is said to mean "the meat-eater that overcomes its prey in a single bound." This much information in a seven-letter word would surely be a masterpiece of abbreviation, yet the description is clear. We cannot doubt that the animal is a versatile killer, for it feeds on capybaras, wild pigs (peccaries), smaller mammals, turtles, large lizards and snakes, crocodilians, big freshwater fishes, and whatever domesticated animals it can poach from the ranches and farmlands that adjoin its shrubby and forested haunts.

Almost nothing is known about the courtship and mating habits of jaguars in the wild beyond the fact that breeding occurs in January and early February at both ends of the long geographic range: Paraguay in the south, and along the United States–Mexico border. In the tropics, where jaguars rarely venture higher than 8,000 feet elevation even at the equator, the mating season seems less definite; kits are found in and around jaguar dens at almost any time of year. Elsewhere, the 100-day gestation generally ends in May with two to four kits, each well furred but with its eyes closed like those of a newborn house cat. Unlike a tomcat, however, the father jaguar stays close by and protects the mother with her young. When the kits are weaned, he helps his mate for a while in training the youngsters to hunt and take care of themselves. Generally he wanders off at the end of the year, whereas the mother continues her parental leadership until she is almost ready to give birth to a new litter. The two-year-olds must go off on their own. They will soon be sexually mature and, with luck, have another twenty years ahead of them.

We naturally compare the jaguar *(Panthera onca)* with the extremely similar leopard *(P. pardus)* of Africa and southern Asia, especially since both of these big cats are marked with patterns resembling cat footprints in black on a tawny yellow coat, or are all black (in South America and Asia). Many of the spot rosettes on a jaguar have a black dot in the middle, whereas those on a leopard do not. A black leopard can be distinguished by its slimmer body, a maximum weight of 200 instead of 300 pounds, a tail three feet long instead of two feet long, and a distinctly smaller head. The leopard always breeds in January, whereas the other two members of the genus—the tiger *(P. pardus)* of the Himalayas and southern Asia and the lion *(P. leo)* of western India and Africa south of the great deserts— can mate at any time of year.

North America's Wildcat

Discovering a wildcat before it vanishes is like a pat on the back for fine woodsmanship. These bobtailed animals are so elusive and nocturnal that they are generally overlooked. They range across southern Canada from British Columbia to Nova Scotia, south to the tip of Florida and into Mexico to the Isthmus of Tehuantepec and the end of Baja California. Only the corn belt states seem to offer a wildcat too few amenities to allow residence.

We find the tracks of wildcats in deep snow each winter, and occasionally a tree where the animal has scratched to sharpen its claws. Each January, if we are lucky, we hear a prowling male. Weighing perhaps 40 pounds and measuring 40 inches from the tip of his short nose to the base of his six-inch tail, he strides along while yowling the most spine-tingling calls of the year. Sometimes a second male answers, and the two engage in a duet followed by sounds of battle so loud and ferocious as to make the equivalent among tomcats seem a sham. Apparently a tabby cat in heat can be impressed too, for hybrids between wildcats and the domestic species are known.

In wildcat families the male is promiscuous. He leaves each mate soon after copulation and has nothing to do with the one to four kittens that she will bear about 63 days later. The mother returns to feed her nurslings regularly until they are about two months old, unless the den is discovered; then she may simply abandon it and them. At two months, she weans her youngsters while training them to hunt. At this stage her maternal responsibilities impress her more, for she will stay to defend her family against a dog or a coyote. By fall, the young wildcats weigh about twelve pounds each and become independent. The females among them, at least, will be sexually mature by the end of the year.

A cornered wildcat puts up a desperate fight, which gives real meaning to the boast of the American poet Eugene Field that his frontier hero "could whip his weight in wildcats." That would be four or five wildcats at once! The same energy is available to subdue a prey animal upon which a wildcat has pounced. The variety of fare is far more varied than it is for the Canada lynx, which is the wildcat's northern kin. A wildcat can find a quarter to a half pound of meat each day almost anywhere in a forest, a field, or a desert by pouncing on cottontails, jackrabbits, unwary rodents (including squirrels), and birds. Stomach contents show that birds are hardest to catch, for they account for only about 3 percent of the diet. Plant foods often amount to 16 percent. More than two-thirds of the remainder is the meat of mammals that man regards as harmful to his interests.

In the Arizona desert, we have watched a wildcat stepping silently along in late afternoon, apparently too hungry to wait for darkness before setting out to hunt. The animal paused to sniff at each rodent burrow, going to one after another by sight and revealing to us by a switching of the short, crossbarred tail when the doorway offered an interesting odor. In the slanting sunlight that brightened the reddish brown fur, we recognized why the wildcat is called bay lynx as well as bobcat. Its tail showed clearly the white tip that distinguishes the wildcat *(Lynx rufa)* from the forest-dwelling Canada lynx *(L. canadensis)* and its Eurasian counterpart the European lynx *(L. lynx)*. But until the daylight faded, the wildcat caught no prey. The gravelly soil between the harsh shrubs was still too warm and the air too dry for the lesser desert dwellers to venture forth and yield the cat a meal.

The Resilient Coyote

The deserts in the American Southwest have given us many a glimpse of a coyote *(Canis latrans)* or two, and in the evenings we have sat enchanted after dark listening to the yelps and songlike falsetto howls of these wild dogs as they called to one another across a valley. The coyote songs seem bright and melodious, not mournful like the notes produced on similar occasions by their nearest kin among dog kinds—the slightly smaller jackals of the Old World.

Until western ranchers began an intensive campaign to kill off the coyotes, these animals were seldom seen east of the Great Plains. Then, as now, they ranged north to Alaska and south to Costa Rica, right to the Pacific coast. But when uncounted numbers, totaling in the millions, were shot and trapped and poisoned in the West, the coyote spread eastward to Florida, New York State, and Ontario, where no comparable campaigns were mounted. Until then, few scientists had taken the trouble to learn what coyotes eat. Stomach contents prove to be about one-third rabbit meat, one-fourth carrion, and nearly one-fifth rodents that do damage to agriculture; of the remainder, about an eighth was meat of sheep or goats, most of it from dead bodies the coyotes scavenged. In winter, the proportion of carrion rises and includes some from dead deer, while the amount of meat from domesticated animals drops to around a twelfth. Coyotes are almost the only predators left in America that can catch and kill jackrabbits as a major source of food.

An individual coyote generally remains on familiar ground, in a territory averaging about six square miles. Usually it is kept well marked by scent posts that are wetted with urine every time the animal comes by. Except in breeding season, which begins in January, the boundaries are normally observed and respected. But when an unattached male goes courting, he ignores the signs and is willing to test his strength against any incumbent male. Strangely, the female may reject the victor if the vanquished one has been her mate for a year or more. She seems to remember that he was the provider when she was pregnant or guarding young the last time. The father usually works hard to bring food to his mate and the pups, and to help teach his young to hunt. He will also serve as a wily decoy to lead dogs or men away from the den if the mother does not try this trick herself before he discovers danger in the vicinity.

There seems to be no rule about which den, his or hers, is to serve as the nursery. Often it is an old badger hole that the coyotes have enlarged. A small cave in a mountain side or a space under a pile of rocks will suit the parents just as well. They live together in it for about two months after the female has come into heat and been inseminated. During her pregnancy, their courtship continues with affectionate nuzzlings and soft whimpers. This behavior reinforces the pair bond, and keeps the male interested and close by even after his mate gently expels him. To give birth and while suckling her pups, she insists on being alone in the den. He must find shelter elsewhere for about two more months. Then he is welcomed back and given a chance to meet his offspring.

When the pups reach five or six weeks of age, they are allowed to sun themselves and play at the entrance to the den. Their ears are bigger and their eyes smaller than those of a dog at a comparable age. Not until they have two adults looking out for them, however, are they led away on hunting trips and shown how to catch mice or rabbits. At first, the youngsters are awkward, excited, noisy, and completely unable to surprise prey or to kill it efficiently. By the time they are experts, autumn has arrived and their parents are rejecting them. They can no longer stay within their parents' territory, but must seek a place of their own. At the risk of their lives, they disperse. Some find a home more than a hundred miles from where they started out. By the first of the next year, the survivors are mature.

To most people, a coyote is just a gray shape running across a field or a road, leaving tracks like those of a dog. The animal holds its ears up and forward, its tail low and drooping—not high, as does the much larger wolf. By learning quickly and showing extraordinary skill in adapting to new situations, the coyotes have increased their range and their numbers despite efforts to get rid of them. In many states they have already made friends with local dogs, giving rise to hybrid "coy-dogs" in which are combined some of the cunning and most of the scavenging habits of the coyote ancestor. That there are not more of the hybrids seems due to the fact that the male coyote is rarely interested in mating before December or after mid-March. His inherited schedule for reproduction is completely adequate in the family life of coyotes, since the females of his species come into heat only early in the year.

The Gray Whales of the North Pacific

To count the gray whales *(Eschrichtius glaucus)* as they pass along the California coast at San Diego, little groups of scientists and laymen set up spotting telescopes on tripods almost daily for about two weeks twice each year. These people gather at a high point in a public park that overlooks the Pacific Ocean. The first part of their whale watch comes late in December, without even an interruption on Christmas Day. The spectacle is more impressive then than when the whales migrate northbound around the first of March, because the great animals swim closer to shore, spouting and slapping their tail flukes on the water, but traveling with evident haste. On their return trip, the whales spread out more and take time to dive for food. They tend to dally, crossing one another's routes so often that it is difficult to keep count.

The whale watchers are determined to see for themselves whether the protection that was extended in 1938 by the International Whaling Commission and the United States government is having the desired results. A century earlier, the whalers operating in Pacific waters killed so many of these 30- to 50-foot whales that, for years, none at all were seen passing San Diego in either direction. By the winter of 1952–53, the number rose to 2,894 counted on the southbound trip. In 1956–57 the tally showed 4,454, and since the middle 1960s about 8,000. So long as protection is maintained, these whales seem to be in no danger of extermination.

On their southbound migration, the gray whales produce no detectable underwater sounds that might serve either for finding food or for gauging the distance to the bottom or to shore. Probably the animals are fasting, and will eat no more until they are traveling northbound again. They stay mostly within two and a half miles of shore and appear to look for land-

marks along the route by rearing up out of the water and then sliding back in. Whale watchers call this behavior "spy-hopping." Yet it does not prevent some gray whales from being stranded on sand bars and beaches for several hours at low tide. Grays are virtually unique in being un-harmed by brief exposure to air and some loss of buoyant support. Nor do they show signs of panic while immobilized if they are approached (but not touched) by people in small boats.

Anyone close to so huge a beast marvels at its bulk. Visitors to a gray whale also discover that its color is unnatural. Its skin is actually black, but blotched extensively and discolored by barnacles and other affixed forms of life. At intervals, the monster exhales forcefully through the paired blowholes atop its head, emitting a jet of oily vapor ten feet high. At close range it is sometimes possible to see into the whale's mouth and glimpse the strange straw-yellow plates of whalebone that hang down on each side, at approximately right angles to the long axis of the animal.

The destination for the southbound migrants is the Pacific side of Baja California, where several extensive shallow lagoons open. Best known are the ones called Vicaino-Scammon, San Ignacio, and Magdalena. During January, the pregnant cow whales hurry into the lagoons and push on for 20 or 30 miles from the entrance. The warm salty water there is the end of their 6,000-mile journey from the icy North Pacific, and the place where the whale ends her eleven-and-a-half-month pregnancy. While giving birth and for about two weeks afterward, the mother lies on her side or her belly, fasting but suckling her huge calf. It is 16 to 19 feet long, and weighs around 3,500 pounds. The mother will be average if she is 38 feet long and weighs a ton for every foot.

Meanwhile the cow whales that have just reached sexual maturity and those that have recently weaned the calf they bore the year previously are sporting in the open ocean close to the entrance to the lagoon. The gray bulls put on exhibitions of fancy swimming, leaping from the water to crash back in, standing on their heads with their tail flukes waving in air, and producing a tremendous uproar of underwater sounds. At intervals, a cow responds to this community-style courtship in some way that indi-cates her willingness. Almost instantly, two bulls swim to her, one on each side, and accompany her into the lagoon to the nearest shallow water. There one bull overtakes her and forces her down until she is supported by the bottom. He mates with her for several minutes, while the second male presses close, apparently to stabilize the pair. Afterward, all three whales swim back to open water. The cow goes off, but the bulls return to their display.

Until recently these antics in the remote lagoons of Mexico remained a mystery because the whales would not tolerate men in small boats approaching the mating and calving areas. But now the scientists can hover in a helicopter, observing and photographing the whales without

Gray Whale (*Eschrichtius glaucus*)

disturbing them in the least. To keep from ruffling the water with the downdraft from the helicopter blades, the voyeurs must maintain a fair distance and compensate for altitude by using binoculars and telephoto lenses. Seemingly the roar of the helicopter interfers in no way with the underwater communications of the gray whales.

In the western world, nothing is known about the present state of a second population of gray whales that have long migrated along the Asiatic coast between the Okhotsk Sea in summer and Korean waters or perhaps those of the Yellow Sea. They do not mix, so far as has been discovered, with the American population. Their calving lagoons and breeding places are probably as similar as geography allows, and their inherited schedule essentially identical. They had an opportunity to recover from exploitation by whalers when military ships and aircraft dominated the Pacific Ocean during World War II. But whether the Asiatic gray whales have been left to follow their ancient ways, as the International Whaling Commission has ordered, remains a secret.

CHAPTER TWELVE

While the Groundhog Sleeps

ON GROUNDHOG DAY, which New Englanders have been observing on February 2 ever since colonial times, a good many kinds of mammals all over the world are responding to the change in length of day by showing a fresh awareness for members of the opposite sex. The American groundhog is not among them, for it is still fast asleep in winter dormancy.

From our own experience with February in New England, we know how much the colonists needed an entertaining fable to cheer them up at the beginning of this month. Since they knew so little about the animals of the New World, they simply adopted a story from northern Germany about a burrow-making mammal with small ears, short legs and tail, and grizzled fur. According to the story, it came out of hibernation on February 2, looked for its own shadow, and seeing this evidence of the sun, went below ground again for another six weeks of winter.

Some of the colonists may have known that the tale referred to the Old World badger *(Meles meles)*. Since they found no badgers in New England or along the Atlantic coast, they transferred the tradition to the groundhog—a burrow maker only a third as big as the European badger, a vegetarian instead of a carnivore, a plain animal instead of one marked with three lengthwise snow-white stripes upon its head. Happy Groundhog Day! The story appealed to wry colonial humor.

Now that the tradition has been traced to its source, we have reason to wonder about the Old World badger in relation to the weather in early February. The animal hibernates until March from Scandinavia across Poland into northern Russia, but remains active all winter farther south. Along the border line the animal sleeps more lightly. It might come out in early February to observe the sun, and then go back into its den for six weeks more. The Eurasian badger mates in summer, between July and September, as does the North American badger *(Taxidea taxus)*, which we meet in states from Ohio west and from the Canadian prairie provinces to Mexico. Yet in both species, although the fertilized egg develops for a total of only about six weeks before becoming a full-term fetus ready to be born, the actual births occur in March or April, earlier in the southern part of each range and later farther north. It is as though mating had occurred around Groundhog Day.

The paradox arises because a badger embryo does not continue its development in the way that the great majority of mammals do. Within a week after fertilization, the embryo pauses in its growth at a stage called the blastocyst. It waits in the womb for a signal of some kind, which scientists have not yet fully identified. Suddenly, in the winter season, the embryo responds by resuming activity. It burrows into the wall tissues of the womb and quickly completes its development, readying itself for birth.

In a pregnant badger that is not hibernating, perhaps in southern Europe or North America, the renewal of embryonic growth may be triggered by an unthinking reaction of the mother to the angle of the sun at noon and the changing length of day. A pregnant badger, hibernating in northern Germany, may well get the same cue in early February by coming to the doorway of her den, not to see her shadow but to let the sun reset her inner clock. If, in consequence, her blastocysts implant themselves and proceed toward birth, the renewal of activity in her womb erases the history of delay. It is as though she had mated just the week before.

Until the British zoologist G. N. D. Hamlett compiled the evidence into a review report in 1935, no one realized how many mammals of different families have exceptional representatives in which the embryos show this phenomenon of delayed implantation. It was suspected first because the time between mating and birth in some species seemed so variable and often so outrageously long in relation to the degree of development in the young at birth. Then scientists began to recognize delayed implantation and the adaptive value in allowing these particular kinds of mammals to mate at a time that was convenient for the adults, and to give birth very much later in a season that was advantageous for the newborn.

Some pouched mammals and armadillos and many bears, weasels and their kin, seals, and roe deer make regular use of the dormant stage in embryonic development. The gray seals that routinely give birth on the ice, or on frigid ledges and sandbars along the Canadian east coast in January and February, mate again almost immediately. But their embryos soon pause in the blastocyst stage and wait about six months before going on. They implant at just about the time that gray seals are mating in the eastern North Atlantic.

At least one of the tropical fruit bats in the Old World shows delayed implantation. Most female bats in the North Temperate Zone merely maintain the sperm alive long after mating has occurred, until they release their eggs in February. Their actual pregnancies begin while the groundhog sleeps, as though this were breeding season. Yet the two sexes get together in late summer or autumn while the bats are alert and sheltering in proximity.

It seems most incongruous for the eggs of bats to be fertilized on Groundhog Day (or some date within a week or so) in the hibernating types, since the male, at least, is utterly unresponsive, deep in dormancy. Despite similar inactivity of the female parent, her inner clock silently signals the time for her ovaries to awaken. They release eggs to meet the sperm as though she were alert to the cyclic changes in the duration of daylight and the height of the noon sun above the horizon. The pipistrelle *(Pipistrellus)*, which is the smallest European bat, combines in its womb the sperm it has maintained from an autumn mating with more from a second breeding season in early spring, all to produce a single brood in June or July.

The Original Owners of Mink Coats

Even in the most northern part of its range, far up on the polar tundras, a slight increase in day length lets a male mink in North America *(Mustela vison)* or Eurasia *(M. lutreola)* know when February arrives. This month and March are his season for visiting, although any young he fathers will not arrive until after about seven weeks of delayed implantation and six weeks more of actual gestation. He slithers out of the shelter where he has been sleeping for a large proportion of each 24 hours, curled up in his warm, waterproof coat, with his eight-inch tail wrapped around him. His hideaway is seldom far from water. It may be a natural cavity in the bank of some stream, or a foxhole, or a rabbit warren, or a den amid a pile of rocks. In the New World it is likely to be a muskrat house, whose owners he has killed and eaten. The mates he seeks will be in similar sites.

The fur of the mink remains dark brown all winter, and contrasts with whatever snow he must cross to reach his many destinations. He risks being pounced upon by an alert owl or a hungry bobcat or fox. Benefitting from the darkness of the long night, he must still be cautious as he enters a doorway where he smells a female of his kind. In his excitement, he exudes musk from a pair of glands beside his anus. The stench offends a human nose, and seems stronger the bigger and older the male mink is, yet it tranquilizes a potential mate. Instead of darting at him with sharp teeth ready to tear at his flesh, she makes good on the promise that her own perfume has given. She does not release it until her eggs are ready for fertilization. Her body and her nervous responses are all attuned to the time of year.

Early in the annual mating season, the patience of female mink is often brief. After she has been impregnated, she repels the male. His sex drive is especially strong at first, yet he needs only a slight rebuff to send him on his way toward another conquest. By March he generally finds a female whose tolerance for him matches his slowing pace. He settles in with her, and stays until late summer, helping in parental chores until the young are independent.

A litter of mink may be as small as two or as large as nine; three to five is an average family. When newborn, their eyes are sealed shut and their bodies covered only by fine short hairs of a pale gray or yellowish color. Their mother starts to wean them by bringing them meat before their eyes open at five weeks of age. As soon as they can see, she gives them no more milk. At six to eight weeks after their birth, they will be old enough to follow their parents out of the den on hunting expeditions. It is time for them to suspend the playful quarreling with which they have filled their hours in the nursery, and to learn the important techniques needed to catch fish, frogs, young turtles, marsh birds, mice, and the larger water insects. Before autumn, each young mink must be able to kill a big musk-rat without help, and to go off on its own to claim a territory.

For an adult female mink, 20 acres along a stream or pond is sufficient space. Within it she will find many hideaways and travel from one to another at random, using each for a few days as a base from which to make nightly forays for food. A male may range over 100 acres, leaving his scent on his own dens. Like a female, he prefers to drag home what bodies he cannot eat where he makes his kills. Soon each den is littered with partly devoured trophies and the skeletons of those that have proved especially tasty.

Wherever mink escape being trapped for their pelts, they build up their populations rather rapidly. Yet, as active predators, their numbers can never become very large because fresh meat becomes hard to find each winter. Even if actual starvation does not cause mink to die prematurely, they become thin and lose some of their natural resistance to fatal diseases. In addition, their frequent encounters as they search for food over wider areas affect the hormone balance in the bloodstream of the pregnant females. Merely seeing other mink is enough to irritate these bearers of the next generation. They secrete extra amounts of adrenal hormone, and this suppresses the production of the normal sex hormones needed for proper implantation of the blastocysts and development of the embryos. Fewer embryos implant and more die instead of growing to full term and being born. The mink population levels off automatically to match the amount of food available. By the time the next generation is born and weaned, they will find plenty of prey because the amenities of summer lie ahead.

The Playful Otters

Almost anywhere a mink can live, a river otter *(Lutra canadensis)* is likely to raise a family. It prefers to be close to a good-sized waterway, where it can excavate a den in a sloping bank. The animal has no aversion to taking over the burrow of a muskrat or a bank beaver, especially if it is close to a place where the otter can go tobogganing—down a mud slide in summer or a snow slope in winter. It is a playful animal at any age and all times of year. If the contours of the land provide no good river banks and food is plentiful the otter will settle for a marsh. In northern California and southern Oregon we have found the homes they have constructed out of the bulrush known as tule grass *(Scirpus).* They pull together the upright leaves to form a sort of wigwam with a spacious central chamber.

Unlike a mink, which is slender like a weasel, an otter in good condition is always rather roly-poly. A thick layer of fat under the skin insulates the otter from even the coldest water and serves also as a reserve of food. It gives the animal an outline suggesting that of a seal, streamlining it for underwater activity. Even the tail, which may be 19 inches long on a 55-inch body, tapers from a thick base to a pointed tip. The hind feet have webbing between the toes, which is unusual among members of the weasel family. They propel the animal so rapidly that it has no trouble catching up with fishes and seizing them in its sharp teeth. It snaps up crayfish at all times of year, and extends its meat diet in summer to include water beetles, frogs, and snakes. While hunting for these elusive morsels, an otter can hold its breath for almost fifteen minutes. Yet in season it is ready to fill its stomach with ripe blueberries and other sweet fruits that grow in otter territory.

We rarely see a solitary otter except in February and March, when the big males go courting. The youngest of them are two years old and already weigh more than fifteen pounds. If possible they swim below the ice to places where a spring keeps it thin, or warm enough to maintain a small pool of open water. Out comes the otter, to shake himself vigorously before his wet fur freezes. He may roll in the snow to finish the drying operation. Then away he bounds along the river bank, hunting for the den in which a ready female may be waiting.

Both sound and scent probably tell him her location and her condition, for his aim is to visit her soon after she has given birth to her annual litter. Her babies—usually two or three—will be whimpering a little as they keep pressed to their mother for warmth and nourishment. Their eyes will be closed for about five weeks, and their thick short fur is so dark that they seem to be mere plump shadows in the den, each somewhat larger than a full-grown chipmunk. Their mother stays between them and the male, but she lets him nuzzle her and eventually copulate. Only her little family is strictly off-limits for him. After all, he just might be hungry enough to eat a baby otter.

After the male leaves, the mother is alone to look after her youngsters. It is entirely up to her to decide when they can be led from the den and taught to swim and hunt. After she has weaned them, she may join forces with another mother otter with a similar family. Through spring and summer and autumn, they are likely to stay together. Young and old play for hours. They comb each other's fur, starting at the head and continuing to the tip of the tail. They dive for food, often coming up with snails or freshwater mussels whose shells they crunch to get at the meat inside.

All this while, for a total of nine months or more, the embryos from the winter mating are still dormant as blastocysts in the womb. Not until late fall or early winter will they resume their development. Their growth must bring them to readiness for birth on schedule, almost twelve months after the parents bred. By then the mother otter will have sent off the youngsters of her previous litter to seek their own fortunes. Her year is about to repeat with a new family and a new mate.

Curiously, the year-long pregnancy is a pattern of development that seems confined to the river otters of North America *(Lutra canadensis)*. In Europe, the corresponding animal *(L. lutra)* mates in late February or early March, but its young develop continuously for about two months, to be born in May. No one knows whether one way is better than the other. The existence of two working systems may be just an example of the rule that the American mammalogist Paul D. Errington recognized: "Nature's way is any way that works!"

Skunks with Spots and Skunks with Stripes

The most distinctive members of the weasel family in the New World are the skunks, of at least ten varieties. Sometimes Europeans mistakenly refer to them as polecats, largely because the polecats *(Mustela putorius* and others) of the Old World are noted for the malodorous secretion they emit from anal glands when defending themselves. A skunk can do far more: it can squirt twin jets of the secretion, combine them into a single stream, and strike a target ten feet away or more time after time.

Striped skunk *(Mephitis mephitis)*

A skunk would rather retire at a reasonable pace than release a stench. Yet if challenged, it turns and makes a stand. The little spotted skunks *(Spilogale)* even perform acrobatic hand-stands, as though to make sure that their contrasty pattern in black and white fur will be seen and recognized. The larger striped skunks *(Mephitis)* turn sidewise and raise their conspicuous white-striped black tail plumes. Most larger animals learn to veer away without pressing the confrontation. Only the great horned owls and eagles, which lack a sense of smell, and hungry badgers, bobcats, and coyotes, which will ignore the odor to get a meal, press the attack and kill the skunk. Other skunks die of diseases, after being weakened by internal parasites. Or they cross a busy highway at their characteristically deliberate pace and are run over by speeding automobiles. Less than a third of the skunks that are born live to be a year old and sexually mature.

Despite the deaths on the highways, there may be more skunks in North America today than when the first European settlers arrived. The felling of the forests and the destruction of the larger predators both favored the skunks, for they find more insects, mice, fruits, and other foods in open areas than under the tall trees. Even a disused space below a porch or in an abandoned shed can serve as a shelter for a skunk family, in place of a groundhog hole or a fox den or some burrow the mother herself might dig. The skunks emerge from these hideaways soon after sunset to do their foraging in darkness. So long as the ground remains unfrozen and no thick blanket of snow keeps them from their food, they may be active all winter long. In bad weather, sometimes for week after week, the skunks may curl up and sleep as though to ignore their growing hunger.

Winter is the most sociable time of year for the sleepy skunks. Several males may den together. Or a male with several females. Or two females and their young of the preceding summer. An example can be found of almost every possible combination. Thus when the adult males begin mate hunting about the middle of February, they never know what they will find as they nose into some strange sleeping quarters. The resident male may protect the females he is keeping company by dousing the newcomer (and the shelter) with scent. Or he may come outside and do the job with less indoor commotion. Whether, upon returning, the male consummates a union with each of his female companions is not known. The most obvious fact is that older females become receptive to a mate several weeks before the younger females do. This spreads the breeding season for the males over March and sometimes into April. On their self-appointed rounds, the males sometimes travel five miles in a night and visit half a dozen females along the way.

East of the Rocky Mountains, a pregnant skunk ordinarily gives birth about 51 days after mating. If she is young, her little family is likely to include only two or three kittens. An older and larger mother may have eight or more. For six to seven weeks, the young will nurse before they begin to sample the foods they see their mother eating while they follow her on her nocturnal expeditions. Before the summer ends, the kittens should be at least half grown and ready to become independent.

West of the Rocky Mountains, the spotted skunks are more likely to postpone the birth of their babies by five months of delayed implantation. This brings a peak of births in October, weaning in December, and independence before the challenging droughts of the next summer make food scarcer than at any other time of year.

In the south-central states and Mexico, the spotted skunks produce young at two different times of year. Some mothers may have two litters in succession; but it is also possible that females give birth in the spring or the fall according to whether they have nurtured the young in their wombs for continuous development or have held the embryos in the blastocyst stage until the hot dry summer was almost over. We cannot think of this as voluntary; instead, it appears to be an on-going test within the pattern of variability of living things, toward discovering which among several possible programs would best meet the challenges from the environment.

The Foxy Ones

Coexistence never comes easy in the wild, any more than among human societies. The limited resources must always be shared among several forms of life, each of which is superbly adapted to exploit its world. To compete successfully, every type of animal needs to be a specialist in some way. Yet for predators to specialize is especially hazardous, because the prey they seek may fluctuate widely in abundance. Unless a predator is prepared to substitute plant foods for prey at some times of year, it is always at least two steps removed from the most regular source of energy for existence—the sun.

It can be no accident that the meat-eaters come in several different sets, each with its characteristic body form and habits in hunting. The dog set in family Canidae, the cat set in family Felidae, the weasel set in family Mustelidae, show differences that are readily recognizable. Within each set we notice a range in sizes both of the predators and of their assortment of prey. The wolf *(Canis lupus)*, weighing up to 80 pounds if female and up to 150 pounds if male, chooses for itself and its young larger prey than the coyote *(C. latrans)* weighing no more than 50 pounds. Quite different diet serves the red foxes *(Vulpes)* and gray foxes *(Urocyon)*, which weigh to 12 or 14 pounds, and the little arctic fox *(Alopex lagopus)*, which matures at between 6 and 12 pounds. Although the arctic fox gains by being an efficient predator on 2-ounce lemmings, it frequently loses by being the prey of a neighboring wolf, which might ignore a lemming in the snow.

The arctic foxes go hungry when lemmings become scarce, as they do almost every fourth year in the Far North. Unlike the snowy owls, which are their chief competitors for lemmings, the foxes cannot fly south to hunt for prey. But in bad years, every male that has failed to find a mate may set out southward on foot. Some venture as far as the northern coniferous forests (taiga) across Eurasia and in North America to the nearest fringe of the Canadian prairie provinces and areas of Labrador that are south of their normal range. Their departure relieves somewhat the danger that the pregnant females and attendant males on the home territory will starve before spring. But whether the emigrants ever find their way north again is still unknown. The lucky ones see a polar bear, and follow it along a coast—even out onto an ice floe twenty miles from shore. Since the big bear eats only about half of each seal it catches, the fox can feast on the remainders.

About 52 days after her February mating, a female arctic fox gives birth to from six to a dozen young. A few days before they are due, she and her mate find a branching den to renovate. Digging a new one in the frozen ground is almost impossible. A few pairs are less resourceful. Then the mother is seen rushing about amid the snow drifts of May, carrying in her mouth a shivering newborn pup or two as she searches for suitable shelter. Without it the pups, clad only in thin coats of brown fuzzy fur, will never survive the polar winds. Not until autumn will they possess the dense woolly coat that is characteristic of arctic foxes, nor the pad of long hairs that keep the feet insulated from the snow.

A good den is no guarantee for a young fox. Each one will have to fight for a place to nurse if the number of pups is greater than the number of nipples. A weakling is shoved aside and soon starves to death. Fighting continues when the parents begin the weaning process, bringing home carrion for the youngsters. No more than half of the litter may survive until the time the milk supply tapers off. Half of the remainder usually perish within a year of birth. The number of two-year-olds is rarely 10 percent of the total that the mothers contributed to that particular year's newborns.

All summer, the arctic foxes find more to eat. They perfect their hunting skills on migrant birds as well as lemmings. Along the coast, the foxes devour the eggs and young of countless gulls, terns, shorebirds, ducks, and geese. The predators invade the nest burrows of puffins and auklets, eating what they can and hauling the remains back to their own dens. Inland, the foxes hunt for nests of ptarmigans.

Often a favored nesting area for waterfowl becomes pocked by new dens dug by arctic foxes, which need places to hide from golden eagles, wolves, and bears, but want to be close to a food supply. The one territory the foxes avoid is close to any nest of a snowy owl. The white owl is alert to drive foxes away, and also catches every lemming that comes within sight, leaving none for a fox. Some of the waterfowl, particularly the snow geese, black brant, and common eider, find sanctuary from the foxes by nesting close to a snowy owl, from which they have little to fear.

From May until October, the foxes eat well and get fat. After the migrant birds depart in early autumn, the tundra still offers many ripe fruits that foxes enjoy. The change toward a simpler, winter world shows on each fox as it sheds its gray or brown summer fur and exposes the thick coat that is growing in. A majority of the foxes will be all white. But a number will be slate-blue except for brownish tail and feet. On the Canadian mainland and across northern Eurasia, less than 1 percent of the arctic foxes inherit the blue pelage for winter. On Baffin Land, the proportion may be as high as one blue in twenty; and in western Greenland, the white foxes barely outnumber the blues. During the dark winter, both hues seem equal in protective value. After spring equinox, the reverse change is under way. The browns and grays reappear first on the tail, and progress farther forward as the nights shorten toward another time of plenty.

In the southern portions of their normal range, the arctic foxes must do their best to avoid the northernmost of the red foxes *(Vulpes)*, which are the commonest wild predators of moderate size both in the Eurasian forests and fields and in North America. The American species *(V. fulva)* comes in two phases: rusty red to golden brown above, whitish below, with black only on the legs; or all black except for the white tail tip that is characteristic of all red foxes. A red fox has pointed ears instead of rounded ones, and grows big enough to be able to make a meal of almost any arctic fox. Although the red fox is an expert at pouncing on small rodents, it spreads its impact in the north country by catching hares and rabbits. For these it has no competition from the arctic foxes, but plenty from the wolves, lynxes, bobcats, and western coyotes. Any of these slightly larger predators would kill and eat a red fox if they could catch one.

February seems to be the preferred month for red foxes to begin courtship. The dog foxes (young males) get into many bitter fights at this season to settle which few among them will be dominant enough to court any vixen (female) that is not already taken by a still more dominant dog fox. Probably some of the breeding is casual. But most pairs stay together as affectionate twosomes after the dog fox has won the vixen's permission to copulate. He helps her find a nursery den, or takes turns in digging a new one. He brings her choice morsels he has caught, such as squirrels and forest mice, and seems always ready to lead away a domestic dog or other predator that might threaten the pregnant vixen or her young.

The wiliness of the American red fox and of his European counterpart *(V. vulpes)* has earned him a fabulous reputation. Certainly he is an extraordinarily resourceful animal in leading the dogs that are set after him astray. He runs along stone walls and steps carefully in shallow water to make little sound while going upstream or down before crossing, letting the water wash away his scented trail. He is equally deliberate about marking his territory with urine for the benefit of other foxes, so that none will unknowingly invade his hunting grounds.

The gray fox *(Urocyon cinereoargenteus)* has a black-tipped tail and a pepper-and-salt mixture in the soft hairs of its coat. The dark stripe along the top of its back from neck to tail may be less conspicuous as the animal dodges through the shadows of the forest. If suspicious, or just to rest, a gray fox often climbs a tree. It is much less likely than the red fox to cross open fields, even at night when it does most of its hunting.

The range of the gray fox barely reaches into western Ontario and eastern Quebec. Near the Pacific coast, we are likely to see one as far north as Oregon. Along these frontiers, the grays and reds mate at the same season, and we have learned to distinguish the barking calls of the courting males. A gray ordinarily is a silent animal, but in February the dog fox yaps harshly four or five times in succession at long intervals as though inviting a vixen of his kind to reply. A red fox, by contrast, follows an initial yelp with a ululating song to which a mate may respond with a high-pitched *yuh-yuh-yuh-yuh-yuh* that has been described as a scream. In the south, where red foxes do not find living conditions to their liking, the grays mate in January—all the way through Mexico and Central America to western Panama. Perhaps in keeping with its preference for warm climates, the gray fox seems never to cache uneaten carrion in or near its den, whereas a red fox in the northern states and Canada is likely to store a reserve of food in this way at any time of year.

We have the impression that the gray fox is less at home in water than the red. One February, in central Mexico, we noticed several of these animals that appeared to be biding their time while the dry season lowered the level of a lake. Already the depth was less than eight inches between the shore and an island on which white pelicans were nesting. We waded across easily, carrying our cameras, and felt needless apprehension that the foxes would observe us and wade across the next night. Several days later we returned and found the young pelicans still safe, the foxes still watching and waiting. No doubt this was an annual gamble, for with a week or two more of growth on fish regurgitated by the parent pelicans, the young birds would be ready to swim beyond the reach of any fox. By contrast, red foxes learned to swim to Backes Island, some distance from the shore of Lake Athabasca, Saskatchewan. They raided the nesting colonies of pelicans, cormorants, and gulls for the first time in 1969 and destroyed about half of the young birds. In 1970 the number of young pelicans that survived the foxes fell to less than 100, as compared to about 900 in 1969 and 1,800 in the last year of undisturbed nesting. As a bird sanctuary, the island had lost its value.

The Golden Jackals

The place of the silent, elusive gray fox in the New World seems occupied by a larger, noisy, sociable member of the dog family in the Old World area from the Balkans and North Africa across Asia Minor. The golden jackal *(Canis aureus)* is about the same size as an American coyote, but acts as if it were smaller unless it has the reassurance of backing from the rest of its pack. Perhaps its behavior, which is often called cowardly, evolved while lions were numerous north of the great deserts of Africa. Then the jackals could scavenge on scraps from lion kills more easily than compete with the smaller foxes for reptiles and nocturnal rodents.

A golden jackal retains a whole repertoire of actions that are widespread among less social members of its family. It approaches an unfamiliar individual of its kind in the same way gray foxes that we have observed do: with every hair erect to make its bulk seem bigger, and with its nose almost to the ground, its tail raised at about 45 degrees. The lone jackal that is approached swivels to expose its flank—its maximum silhouette—and seems to grow by arching its back. The two come together like the upright part and the top of a letter T, and decide the outcome of the confrontation. Neither seems to look the other in the eye. Scent must reveal whether the two are of unlike sex, and the reproductive state of a female. Yet an inconspicuous signal, such as a curl of the lip or a slight droop of the ears, surely reveals a slight submissiveness of one jackal. Rarely is a fight necessary to settle the order of dominance. Acceptance into the pack, which consists of as many as 30 individuals, is still rarer if the newcomer is a male.

In late January and early February, males try to establish a one-to-one relationship with an attractive female. Each male trots along over the dry soil with a springy gait, but stops at intervals where his sense of smell identifies a special place. After exploring thoroughly an area of a square yard or less, he sprinkles it with urine, then continues on his way. As mating season gets nearer, he takes the trouble to scrape the ground with his front feet and to urinate only on the roughened surface.

The approach of an adult male to a female follows a stereotyped pattern. It begins as a normal confrontation, which animal behaviorists call a T-sequence. The female makes no move to encourage him. Yet the male shows that he has located a potential mate by whimpering and lifting his tail a little higher. He nuzzles her fur and slowly circles her. If she does not object, he lays his head across her back or gently raises a forepaw to stroke her shoulder. She soon tires of this, and wheels to walk or run away. He follows, and repeats his endearments a dozen times a night if she stands still. By day he snoozes close to wherever she chooses to lie down.

The male gets little to eat unless the two discover carrion big enough to give both of them a good meal. He interferes minimally with the hunting of the female, but takes off almost no time to attend to his own hunger. If another male turns up, he must be ready for a test of dominance. Quite often the intruder is larger and probably older. He stays, while the previous companion of the female runs off to find another potential mate and start his courtship over.

As the single week approaches during which the female jackal will be fully receptive to a mate, she responds a little in a curious way. After he has turned aside for a few minutes to go through his routine of marking a territory with urine, she follows closely and anoints with her own contribution each scraped area or tuft of grass—sometimes at the very moment that he lifts his leg. He is not distracted by this attention, for he seems unready to relinquish his premating behavior until he can actually participate in parenthood.

Patiently he repeats his routine, giving the female one invitation after another to respond. He neither tries to hurry her nor, so far as we can see, shows any indication of surprise when at last she reacts by standing beside him head to tail or by becoming the investigative member of a T-sequence. Both positions let her reconnoiter his maleness by nosing his inguinal region. At first he makes no attempt to caress her in a comparable way, and none to mount her. Not for a day or two more will she tolerate such intimacy even for a moment. He must be quick to desist if she turns toward him with mouth open and teeth bared in self-defense.

For fully a week he must be satisfied with foreplay. Actual copulation comes almost as an anticlimax. Although repeated many times each night, it tapers off toward the end of her week of receptivity. The male wanders progressively farther from her side and hunts for food. For a while his role is finished. Until the pups are born, which will not be for somewhat more than another two months, he can regain his strength. He needs it later when he hunts for food to support his mate and their young, while she stays in the den. His efforts provide the livelihood of the little family that inherits his traditions and can carry them on.

The Gloved Wildcat and the House Cat

The social interests of a male dog in a family of pups contrasts completely with the utter disinterest of the average tomcat in the kittens he has sired. A domesticated cat comes by this self-centered behavior quite honestly, for it can be recognized also in the males of the gloved wildcat, one of the Old World felines that supposedly contributed breeding stock to the origin of the household cat. This nocturnal wild prowler is known by many common names: African wildcat, Egyptian cat, Kaffir cat, Indian Desert cat, and gloved cat *(Felis libyca)*. Its paws and lower legs are usually uniform dark gray, giving the animal the appearance of wearing gloves. Distinctive reddish areas on the back of each ear, the tabby markings in narrow dark lines on a paler gray fur, the black on the tail (including the tip), and the size of the animal—slightly larger than an average housecat —all go together whether the gloved cat lives in Africa south of the driest deserts, or across Asia Minor and northern India into Mongolia. On the dark sands of the Kara Kum desert east of the Caspian Sea, it competes with golden jackals and several sizes of foxes for gerbils and numerous other rodents. It shelters if possible in an abandoned den of a Eurasian fox *(V. vulpes)* or of an Old World porcupine.

Like other members of the genus *Felis,* the gloved cat shows a continual state of sexual readiness in the male, and a brief season of eager receptivity (estrus) in the female. Female gloved cats limit their period to the first half of February, each individual coming into heat for just a few days. All the males within smelling distance come hurrying through the night to sere-nade her and to fight for the favor she must confer if she is to have a family. Around her den the caterwauling takes on all of the variations a kindle of domesticated cats can produce, but loses to the starry sky most of the echoes we are used to hearing from city catfights. To this extent the vocal display and battling are quieter in the open.

Again, the gloved cat resembles the domesticated tabby in producing no eggs for the sperm to fertilize until after at least one male has copulated with her. She keeps the sperm alive until they can function. By then the parenthood of the kittens she will bear 62 days later is likely to be so scrambled that no two may have the same father.

The gloved wildcat and the domesticated cat remain perfectly interfer-tile. Inbreeding proceeds at a fair rate in parts of Africa and Asia where the gloved cats live within a mile or two of human communities. Unques-tionably they come so close to benefit from the rats and mice that raid the stores of civilized food and eat the garbage that people discard. Dr. N. B. Todd of the Carnivore Genetics Research Center in Newtonville, Massa-chusetts, suspects that this exploitation of opportunities afforded by human kind was fundamental in the domestication of *Felis catus* in the first place. Gloved wildcat males still show a strong attraction to the scent of a tame female in heat; tame tomcats visit wildcat females in February if they can compete with male wildcats in the open. Yet not all characteris-tics blend among the offspring of these crosses. The kittens born of a household female mated with a male gloved wildcat always seem to resent being handled. They spit and scratch as soon as they can see what is going on, and eventually escape from the human community. The same mother will produce young that remain relaxed when petted at any age if she is mated to a domesticated male.

The flow of inheritance from home to countryside is filtered too. Somehow the independence shown by female house cats regarding the time of year for coming into heat fails to spread among the neighboring gloved wildcats. The household animal remains unsynchronized with any other cat in the house; she comes into heat two (or three) times annually on her own schedule. She takes her suitors at her convenience, although she sheds her fur in spring and fall. Surely this is an inherited independence, a feature that has evolved during about 5,000 years of domestication, one that gained a place among the behavior characteristics carried by the genes in both eggs and sperm. Somehow the gloved wildcats beyond the town limits seem able to absorb and repress such a casual pattern of reproduction. They persist in observing February as the only mating time. So far no one has discovered what happens if a hybrid individual makes a mistake.

Despite the close association with the house cat and the wealth of information gained from careful observation of the gloved wildcat and the lion, the eagerness of these animals to copulate repeatedly in their season mystifies behavioral scientists. Conception rate is low in relation to the number of copulations and estrous periods. Establishment and maintenance of a strong pair bond does not follow. Perhaps the female uses this means to assess the vigor of the male. Possibly she welcomes his attention as a way to compete with other females for reproductive success. These explanations are not necessarily independent or mutually exclusive. The answer might even be relevant for humankind.

The Most Common Monkeys in Africa

In trees above the thornscrub from Kenya south, the most common of Africa's monkeys spend the dark hours above the realm where gloved wildcats prowl and mate. Each vervet monkey waits for daylight before clambering about. Young ones, still dependent on their mothers, surprise us by using their long tails to help hold on to the branches, whereas the cat-sized adults employ their tails only as balancing organs. Apparently this satisfies the rule that monkeys have prehensile tails only in Latin America.

The French common name *guenon* ("frightening one"), like the generic name *Cercopithecus* ("nearly human") seems equally inappropriate for these social little tree climbers. They do grimace and expose their teeth when nervous or irritated, such as when they descend to the ground to feast on fallen fruits where a predator could easily rush into action from concealment. Vervets *(C. aethiops)* explore on the ground in early morning and late afternoon, when the shadows are long, but never venture far from a vertical route to safety up a tree. They are monkeys of the open forest and the forest edge, not of the dense rain forest or of savannas, where the trees are far apart. As the grass grows taller after each rainy season, the vervets spend less time on the ground and more in social interaction on high branches.

These habits probably explain why vervets and other guenons respond to changes in the proportions of night and day more than most monkeys or than any of the apes. Vervets just north of the Equator in Uganda and Kenya become most active sexually on February nights. Six months later, in August, most of the adult females have newborn young. Learning the normal period of gestation more precisely than this is almost impossible in the wild because female guenons have no "sexual skin" on their buttocks or perineal areas to show by changing color the state of the menstrual cycle. The female remains receptive to the males in her troop for almost two weeks at a time, and continues to invite them to copulate with her at intervals even if she is pregnant. Among vervets, in fact, both sexes appear equally interested in sexual activity. From their antics alone, it would be hard to decide that one season is better than another.

In suitable territory, an old male dominates a family group of vervets and keeps them within an area of about 20 acres. This amount of open forest supplies edible buds, fruit, grain, roots, and some attractive insects and lizards for variety; it meets the dietary needs of fewer than a dozen monkeys. On poorer land, the animals search more widely, sometimes over as much as 200 acres. Yet if some local abundance of food attracts several family groups at once and they combine to form a larger troop, the males seem surprisingly tolerant of one another and seldom fight. Later the members of the small groups sort themselves out again, clearly recognizing one another by voice if not by sight.

A mere 300 to 400 miles farther south in Kenya or in Tanzania, the vervets on the opposite side of the equator compensate for a difference in the timing of the rainy season. They produce most of their young in November, which means that their matings in May are the most fruitful. How they manage to delay production of eggs for three months, so that pregnancies begin and end later, remains a secret.

More than an ability to be both regular and flexible must explain why vervets succeed in being the most common monkeys on the whole continent. Recent studies show that both mature males and females are amazingly able to warn others of their group when they sense danger. They utter distinctive calls, which differ between the sexes but are recognized and heeded. One call alerts all neighbors to the proximity of a large predator (such as a leopard). Another is for snakes (such as pythons), still another for flying birds of prey (especially martial eagles), and a variety of sounds are for dangerous primates (such as baboons and people). Responses to these calls are characteristic and are identical when the sounds are tape-recorded and played back at times when no actual danger is present. The leopard call hurries the vervets into trees, the snake call induces them to look down and study the ground, the eagle call gets the vervets to run into the shelter of shrubbery, away from exposed branch tips and open ground, and also to look upward. When this approach to language was noticed and reported on in 1967 by Dr. T. T. Struhsaker of the University of Chicago, he made no claim that the vervets use words, merely that their grunts and chattering conveyed distinctive messages of importance of survival. A tream of scientists from Rockefeller University, led by Dr. Robert M. Seyfarth, confirmed the reality of the alarm system as designating "different classes of external danger" eliciting different responses. Whether the chattering of other monkeys conceals meaningful messages remains to be investigated more thoroughly.

The Only Pouched Mammal Native to North America

No one credits the cat-sized opossum *(Didelphis marsupialis)* with more than the minimum of suitable reactions to survive in its way of life. Yet during historic times it has expanded its range northward for hundreds of miles, mostly by accepting almost any shelter and an endless assortment of animal and vegetable matter as food. Over a north-south range extending thousands of miles, the opossum shuts its eyes and stiffens its muscles as though rigor mortis had set in whenever it is discovered by a dog or other predator. Scarcely breathing, it fails to react when picked up or dropped. Any predator that catches an opossum for practice rather than from hunger is likely to abandon it. Then the apparently dead animal awakens, looks around, and scrambles off. It has more lives in succession than the proverbial house cat.

At latitudes from Washington, D.C., north into southern Canada, opossums have only one litter each year, but they start early. On winter days when the temperature rises above the freezing point, both sexes are abroad foraging. In February the males begin looking for females as well as food and, for a short time, become willing to forego the natural solitary habits of full-grown opossums to pair off. A female that is only a year old and inexperienced in sex may resent the approach of any male, and continually snap at him. Her lips draw back to expose her sharp teeth. She chatters them and growls at him. Since he is smaller than she, he does well to heed her warning. He seems to count on her eventual acceptance of his services. When she does let him mount, he is quick about it and soon hurries off.

Pregnancy for a North American opossum lasts just twelve and a half days. During the last week, the mother-to-be gathers fresh, dry bedding into her den. With her pointed jaws she arranges plant fibers into a wad and pushes them below her abdomen. In this position she can support them by curling her naked, scaly, prehensile tail between her legs like a long slender finger. Holding the materials firmly, she ambles along to her chosen shelter.

The actual birth takes less than an hour. During the process, the mother lies back with her tail extended between her hind legs. One after another, the tiny young emerge from her birth canal. Each is about the size of a honeybee, and so bare that some of its internal organs can be seen through its translucent skin. It shows no eyes or ears, hind legs or tail. Yet it uses its forelegs to haul itself through the tangled curls on its mother's belly to the opening of her pouch. As many as eighteen start off this obstacle race, with no more than thirteen slender nipples waiting in the pouch for the winners. Any babies that miss the pouch or get there after the restaurant is full keep on crawling until they die. The successful ones must still cope with accidents. Fewer than ten usually survive for the month before they first peek out of the pouch. At five weeks they try emerging altogether. At eight weeks they are weaned and may go off on their own. After twelve weeks, their mother gives them no choice about leaving her. In the southern United States she may be getting ready for a second mating. In Latin America this frequently produces the second of three litters in succession.

Opossum (*Didelphis marsupialis*)

Apparently food is no real problem for opossums throughout their range, which extends into South America as far as Argentina. But already these versatile animals may have reached their limit northward. Across Michigan the boundary coincides with a climatic line beyond which winter weather confines opossums for more than 70 days between autumn and spring. No matter how soundly the opossums sleep when the cold and snow prevent them from foraging, their store of fat cannot last longer than this. They may go through the routine of mating in February, and then starve to death before their young have had a chance to see the world.

A Free-Wheeling Style of Family Life

THE ABILITY TO follow one pregnancy with another, regardless of the time of year, is shared by humankind and a good many other kinds of mammals. It provides courtship and birth in every month, as though these particular creatures had freed themselves from seasonal changes in their environment. More likely it is a special degree of independence, allowing them to ignore the fact that each season is a separate living space with its own inventory of active inhabitants. Yet few of these species, other than our own, manage without an inconspicuous mechanism that regulates their population size in relation to the food and space available.

This free-wheeling style of family life shows no obvious relation to either the proportion of the normal lifespan that precedes puberty or the size of the mammal at maturity. Successive pregnancies are normal for the well-fed house mouse, some lemmings of polar lands, the vampire bat of tropical America, the horse and giraffe, the hippopotamus and warthog, the mountain lion and sea otter, the European rabbit, the majority of primates, both living kinds of elephants, and at least two kinds of whales —including the great blue, which grows to become the largest animal the world has ever known.

For each of these species, a satisfactory explanation may be discovered by studying in depth how the pressure of natural selection was shifted— from seasonal amenities as the young reached independence, to a more critical challenge in the normal life history. The essential measure is the number of young surviving to mate and reproduce as their parents did. Any change in behavior or tolerance that increases this number even slightly is a gain of vast importance. Identifying the beneficial step becomes the dream of the scientific investigator.

A fossil discovered recently in the Afar badlands of northeastern Ethiopia has renewed attempts to explain the free-wheeling style of human family life. The incomplete skeleton, named Lucy by her discoverer Donald Johanson, is clearly that of a small primate that walked on two feet about 3.5 million years ago. At that time the forests that had once spread all across Africa were shrinking progressively, requiring the descendants of ancestral apes to scurry across the ground from tree to tree. How, one might ask, could primitive members of the human species improve the likelihood of reaching the next place of relative safety without losing their children along the way? Anthropologist Owen Lovejoy of Kent State University suggests that Lucy and her brood might have made it if she could concentrate on being a watchful, protective parent, a specialist tending the family until the young could manage on their own. Lucy might have been able to devote herself so thoroughly if she had a mate who could be counted on to bring food to her and her family. He might go off repeatedly on hunting trips with other vigorous males, but he would return to the female and young he regarded as his—his to provision and protect.

Lovejoy implies that this monogamous concentration on one female and shared children is an ancient tradition, one arising easily among humankind by having the female offer herself as bait. By making herself attractive and being receptive to sexual interaction at any time, even during early months of pregnancy, she could encourage her male to come back to her, to serve her needs, to help her specialize in tending the family. More of his and her offspring would survive to reproductive age than if she had to combine gathering food for herself and them with carrying and nursing the latest baby, and supervising and teaching the older children. By having the male return to her repeatedly, he might even get to recognize certain young as his, and be willing upon occasion or in emergency to look after them himself. This too could bias the world in favor of his posterity.

No one claims that this strategy was devised and established deliberately. Progressive tendencies in the appropriate direction could accumulate unnoticed and bring the same benefits. The change could precede and favor the later development of a larger head and brain, of tool making, of speech and language, of social patterns we could identify as civilization. Like other inherent and learned behavior it would leave no fossil evidence. Yet it could lead to the actions and attitudes basic in every culture.

The spread of human births throughout the year certainly reduces social strains. No surge of young in a particular month distracts parents simultaneously. Instead the accommodations necessary in care for young can be spaced out and fitted into the broad pattern required by a prolonged childhood. Young of many different ages adjust more smoothly into their essential education. Some can start at summer's end and others after New Year's Day. The use of cultural facilities is smoothed out. Competition for gainful places in the adult world is far less than if most juveniles from a year class had to be integrated within a few weeks.

We can appreciate the benefits in the human species from the free-wheeling style of family life while recognizing the advantages available to members of other kinds of mammals as they schedule their courtship and birthing according to season.

Index